Glencoe McGraw-Hill

Grade 8

Math Triumphs

Book 1: Algebra

D1216610

Authors

Basich Whitney • Brown • Dawson • Gonsalves • Silbey • Vielhaber

McGraw Hill Glencoe

Photo Credits

All coins photographed by United States Mint.
All bills photographed by Michael Houghton/StudiOhio.
Cover Jupiterimages; **vi** CORBIS; **vii** Digital Vision; **viii** C. Borland/Getty Images; **002–003** Rodger Klein/Peter Arnold; **008** Frank & Joyce Burek/Getty Images; **010** George Gojkovich/Getty Images; **015** Felicia Martinez/PhotoEdit; **017** Getty Images; **018** Brand X Pictures; **025** Lee Foster/Alamy; **026** Scott Halleran/Getty Images; **030** Ned Frisk/CORBIS; **031, 032** Alamy; **036** Tyrone Turner/Getty Images; **038** James Steinberg/Photo Researchers Inc.; **039** Jack Hollingsworth/Getty Images; **045** Getty Images; **048–049** David Madison/CORBIS; **049** (cw from top)Jupiterimages, (2)PunchStock, (3)Getty Images; **054** Jose Luis Pelaez/Jupiterimages; **055** G.K. & Vikki Hart/Getty Images; **056** CORBIS; **060** G.K. & Vikki Hart/Getty Images; **062** (t)Ryan McVay/Getty Images, (b)Michael Newman/PhotoEdit; **075** David Sacks/Getty Images; **077** Martin Ruegner/Getty Images; **093** Alamy; **096–097** John Giustina/Getty Images; **102** Getty Images; **104** CORBIS; **109** Siede Preis/Getty Images; **116** Jules Frazier/Getty Images; **117** Ken Cavanagh/The McGraw-Hill Companies; **118, 123** Getty Images; **124** Photolibrary; **133** (t)C. Borland/Getty Images; **133** (b)SuperStock.

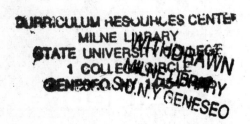
CURRICULUM RESOURCES CENTER
MILNE LIBRARY
STATE UNIVERSITY COLLEGE
1 COLLEGE CIRCLE
GENESEO, N.Y. GENESEO
WITHDRAWN
MILNE LIBRARY

The McGraw·Hill Companies

Macmillan/McGraw-Hill
Glencoe

Copyright © 2009 The McGraw-Hill Companies, Inc. All rights reserved. No part of this publication may be reproduced or distributed in any form or by any means, or stored in a database or retrieval system, without the prior written consent of The McGraw-Hill Companies, Inc., including, but not limited to, network storage or transmission, or broadcast for distance learning.

Send all inquiries to:
Glencoe/McGraw-Hill
8787 Orion Place
Columbus, OH 43240-4027

ISBN: 978-0-07-888213-5
MHID: 0-07-888213-3

Math Triumphs
Grade 8, Book 1

Printed in the United States of America.

4 5 6 7 8 9 10 HSO 17 16 15 14 13 12 11 10

Copyright © Glencoe/McGraw-Hill, a division of The McGraw-Hill Companies, Inc.

Math Triumphs

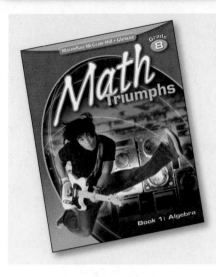

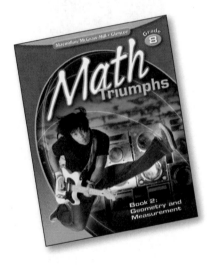

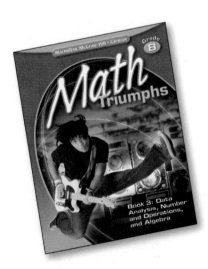

Copyright © Glencoe/McGraw-Hill, a division of The McGraw-Hill Companies, Inc.

Authors and Consultants

AUTHORS

Frances Basich Whitney
Project Director, Mathematics K–12
Santa Cruz County Office of Education
Capitola, California

Kathleen M. Brown
Math Curriculum Staff Developer
Washington Middle School
Long Beach, California

Dixie Dawson
Math Curriculum Leader
Long Beach Unified
Long Beach, California

Philip Gonsalves
Mathematics Coordinator
Alameda County Office of Education
Hayward, California

Robyn Silbey
Math Specialist
Montgomery County Public Schools
Gaithersburg, Maryland

Kathy Vielhaber
Mathematics Consultant
St. Louis, Missouri

CONTRIBUTING AUTHORS

Viken Hovsepian
Professor of Mathematics
Rio Hondo College
Whittier, California

FOLDABLES Study Organizer **Dinah Zike**
Educational Consultant,
Dinah-Might Activities, Inc.
San Antonio, Texas

CONSULTANTS

Assessment

Donna M. Kopenski, Ed.D.
Math Coordinator K–5
City Heights Educational Collaborative
San Diego, California

Instructional Planning and Support

Beatrice Luchin
Mathematics Consultant
League City, Texas

ELL Support and Vocabulary

ReLeah Cossett Lent
Author/Educational Consultant
Alford, Florida

Copyright © Glencoe/McGraw-Hill, a division of The McGraw-Hill Companies, Inc.

Reviewers

Each person below reviewed at least two chapters of the Student Edition, providing feedback and suggestions for improving the effectiveness of the mathematics instruction.

Patricia Allanson
Mathematics Teacher
Deltona Middle School
Deltona, Florida

Debra Allred
Sixth Grade Math Teacher
Wiley Middle School
Leander, Texas

April Chauvette
Secondary Mathematics Facilitator
Leander Independent School District
Leander, Texas

Amy L. Chazarreta
Math Teacher
Wayside Middle School
Fort Worth, Texas

Jeff Denney
Seventh Grade Math Teacher, Mathematics
   Department Chair
Oak Mountain Middle School
Birmingham, Alabama

Franco A. DiPasqua
Director of K-12 Mathematics
West Seneca Central
West Seneca, New York

David E. Ewing
Teacher
Bellview Middle School
Pensacola, Florida

Mark J. Forzley
Eighth Grade Math Teacher
Westmont Junior High School
Westmont, Illinois

Virginia Granstrand Harrell
Education Consultant
Tampa, Florida

Russ Lush
Sixth Grade Math Teacher, Mathematics
   Department Chair
New Augusta - North
Indianapolis, Indiana

Joyce B. McClain
Middle School Math Consultant
Hillsborough County Schools
Tampa, Florida

Suzanne D. Obuchowski
Math Teacher
Proctor School
Topsfield, Massachusetts

Karen L. Reed
Sixth Grade Pre-AP Math
Keller ISD
Keller, Texas

Deborah Todd
Sixth Grade Math Teacher
Francis Bradley Middle School
Huntersville, North Carolina

Susan S. Wesson
Teacher (retired)
Pilot Butte Middle School
Bend, Oregon

Copyright © Glencoe/McGraw-Hill, a division of The McGraw-Hill Companies, Inc.

Contents

Chapter 1 | **Integers**

Navajo Bridge, Colorado

Copyright © Glencoe/McGraw-Hill, a division of The McGraw-Hill Companies, Inc.

Chapter 2

Patterns and Graphs

Saint Louis, Missouri

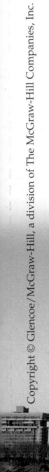

Copyright © Glencoe/McGraw-Hill, a division of The McGraw-Hill Companies, Inc.

Contents

Appalachian Mountains, West Virginia

Copyright © Glencoe/McGraw-Hill, a division of The McGraw-Hill Companies, Inc.

SCAVENGER HUNT

BOOK 1

Let's Get Started

Use the Scavenger Hunt below to learn where things are located in each chapter.

1. What is the title of Lesson 3-2?

2. What is the Key Concept of Lesson 2-3?

3. On what page can you find the vocabulary term *opposites* in Lesson 1-3?

4. What are the vocabulary words for Lesson 2-3?

5. How many Examples are presented in the Chapter 3 Study Guide?

6. What strategy is used in the Step-by-Step Problem-Solving Practice on page 67?

7. List the integers that are mentioned in Exercise 11 on page 24.

8. What is Step 5 in Step-by-Step Practice on page 30?

9. On what pages will you find the Study Guide for Chapter 2?

10. In Chapter 3, find the logo and Internet address that tells you where you can take the Online Readiness Quiz.

Copyright © Glencoe/McGraw-Hill, a division of The McGraw-Hill Companies, Inc.

Integers

Scuba divers measure altitude with integers.

Different marine animals live at specific altitudes. Scuba divers use integers to find different animals by measuring the depth below sea level.

Copyright © Glencoe/McGraw-Hill, a division of The McGraw-Hill Companies, Inc.

Copyright © Glencoe/McGraw-Hill, a division of The McGraw-Hill Companies, Inc.

STEP 1 Quiz

Math Online ⟩ Are you ready for Chapter 1? Take the Online Readiness Quiz at *glencoe.com* to find out.

STEP 2 Preview

Get ready for Chapter 1. Review these skills and compare them with what you will learn in this chapter.

What You Know	What You Will Learn
You know **whole numbers** are **zero** and the counting numbers.	**Lesson 1-1**

What You Know

You know **whole numbers** are **zero** and the counting numbers.

Whole numbers that are **less than** a number are to the left of the number on the number line.

$4 < 9$

Whole numbers that are **greater than** a number are to the right of the number on the number line.

$8 > 2$

To subtract positive numbers on a number line, begin at the *first* number. Move left the *same* number of spaces as the *second* number.

TRY IT!

$12 - 9 =$ _____

The difference is 3.   Begin at 12 and go left 9 spaces.

```
 +--+--+--+--+--+--+--+--+--+--+--+--+--+
 0  1  2  3  4  5  6  7  8  9  10 11 12
```

What You Will Learn

Lesson 1-1

Integers are whole numbers and their opposites.

Opposites are numbers the same distance from zero, but in the opposite direction. For example, the opposite of 3 is −3.

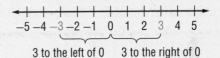

3 to the left of 0   3 to the right of 0

Lesson 1-3

To subtract using negative numbers on a number line, begin at the *first* negative number. Move to the left the same number of spaces as the *second* number.

Example: $-2 - 3 = -5$

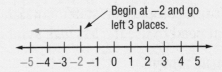

Begin at −2 and go left 3 places.

```
 +--+--+--+--+--+--+--+--+--+--+--+
−5 −4 −3 −2 −1  0  1  2  3  4  5
```

3

Model Integers

Copyright © Glencoe/McGraw-Hill, a division of The McGraw-Hill Companies, Inc.

KEY Concept

Whole numbers are zero and the counting numbers.

Opposites are numbers the same distance from zero, but in the opposite direction. For example, the opposite of 5 is −5.

```
 −5                    5
◄──┼──┼──┼──┼──┼──┼──┼──┼──┼──┼──►
  −5 −4 −3 −2 −1  0  1  2  3  4  5
   └───────────┘  └───────────┘
   5 to the left of 0   5 to the right of 0
```

Integers are whole numbers and their opposites, such as 1 and −1, 15 and −15.

Positive numbers are numbers that are greater than zero.
Negative numbers are numbers that are less than zero.

VOCABULARY

integer
the whole numbers and their opposites
Example: …−3, −2, −1, 0, 1, 2, 3,…

negative number
a number less than zero

opposites
numbers that are the same distance from zero in opposite directions

positive number
a number that is greater than zero

whole numbers
the set of all counting numbers and zero

The number zero is neither positive nor negative.

Example 1

Use <, =, or > to compare −2 and 2.

1. Graph both numbers on the number line.

```
             −2         2
◄──┼──┼──┼──┼──┼──┼──┼──┼──┼──┼──►
  −5 −4 −3 −2 −1  0  1  2  3  4  5
```

2. The number farther to the right is 2, so it is the greater number.

3. Write a comparison statement. Since −2 is less than 2, you need to use the *less than* symbol. **−2 < 2**

YOUR TURN!

Use <, =, or > to compare −5 and 0.

1. Graph both numbers on the number line.

```
◄──┼──┼──┼──┼──┼──┼──┼──┼──┼──┼──►
  −5 −4 −3 −2 −1  0  1  2  3  4  5
```

2. The number farther to the right is _____, so it is the greater number.

3. Write a comparison statement. 0 _____ −5

Example 2

Graph the integers −4, 3, 0, 5, and −2 on a number line. Then write them in order from least to greatest.

1. On the number line, place a dot at each of the numbers.

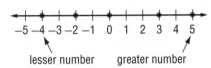

    lesser number   greater number

2. Write the graphed numbers in order as they appear from left to right. **−4, −2, 0, 3, 5**

YOUR TURN!

Graph the integers −3, 2, 1, −5, and −1 on a number line. Then write them in order from least to greatest.

1. On the number line, place a dot at _____.

    −5 −4 −3 −2 −1 0 1 2 3 4 5

2. The numbers in order from least to greatest (as they appear from left to right) are

_____.

Example 3

Write an integer to represent the sentence.

"A reef-building coral is found <u>93 feet below sea level</u>."

1. Underline the key words.

2. Decide if the number is positive or negative.
   negative

   Imagine a number line that is vertical instead of horizontal. Sea level is "0." Below sea level is negative. Above sea level is positive.

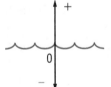

3. Write the integer. **−93**

YOUR TURN!

Write an integer to represent the sentence.

"The temperature in Kansas City one February day is 58°F."

1. Underline key words.

2. Decide if the number is positive or negative. _____

   Imagine a number line that is vertical instead of horizontal. Temperatures above 0°F are positive. Temperatures below 0°F are negative.

3. Write the integer. _____

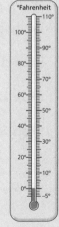

GO ON

Copyright © Glencoe/McGraw-Hill, a division of The McGraw-Hill Companies, Inc.

Who is Correct?

Write −8, 7, 6, and −5 in order from least to greatest.

Omari	Marisol	Renée
−5, 6, 7, −8	−8, −5, 6, 7	−5, −8, 6, 7

Circle correct answers. Cross out incorrect answers.

▶ Guided Practice

Write <, =, or > in each circle to make a true statement.

1 3 ◯ 4

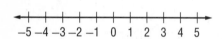

2 −6 ◯ −7

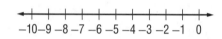

3 −1 ◯ 0

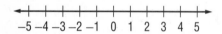

4 −3 ◯ 1 ◯ 3

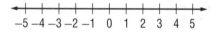

Step by Step Practice

5 Use a number line to compare the integers 6, −3, 10, −8, 1.
Write the integers from least to greatest.

Step 1 Graph the numbers on the number line.

Step 2 What number is farthest to the left? _____

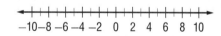

Step 3 Write the numbers in order.

_____

Copyright © Glencoe/McGraw-Hill, a division of The McGraw-Hill Companies, Inc.

Graph the integers on a number line. Then write them in order from least to greatest.

6 5, −2, 1, 4, −4

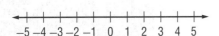

_____

7 3, −5, 2, −2, 4

_____

8 2, −5, 0, 1, −3

_____

9 4, −2, 0, 2, −4

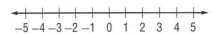

_____

Step (by) Step **Problem-Solving Practice**

Solve.

Problem-Solving Strategies
☑ Draw a diagram.
☐ Use logical reasoning.
☐ Make a table.
☐ Solve a simpler problem.
☐ Work backward.

10 **WEATHER** The temperature in Butte, Montana one morning was 7°F (Fahrenheit). The temperature at noon was 19°F. By evening, the temperature was −8°F. What was the lowest temperature that day?

Understand Read the problem. Write what you know.
The temperature began at _____.
Then it was _____.
By evening, the temperature was _____.

Plan Pick a strategy. One strategy is to draw a diagram.

Make a line to represent a thermometer. Mark the 0. Then mark it in 2-degree increments.

Solve Begin at 7°F. Then mark 19°F and −8°F. The lowest temperature that day was _____.

Check Does the answer make sense? Look over your solution. Did you answer the question?

GO ON

Copyright © Glencoe/McGraw-Hill, a division of The McGraw-Hill Companies, Inc.

11 **SCUBA DIVING** When Isaac went scuba diving he saw a clown fish at 42 feet below the surface. What integer represents this depth?

Check off each step.

_____ Understand: I underlined key words.

_____ Plan: To solve the problem, I will _____.

_____ Solve: The answer is _____.

_____ Check: I checked my answer by _____.

12 **BUDGETS** Martha's Moving, Inc. is $274 below their budget for this month. What integer represents this amount of money?

_____

13 **Reflect** How does the scale on a thermometer help you understand positive and negative integers?

_____

_____

_____

▶ Skills, Concepts, and Problem Solving

Use <, =, or > to compare each pair of numbers.

14 13 and 0

_____

15 0 and −4

_____

16 25 and −25

_____

17 −9 and −7

_____

18 14 and −160

_____

19 −981 and 992

_____

20 9 ◯ 0 ◯ −3

21 −40 ◯ 40 ◯ −4

Copyright © Glencoe/McGraw-Hill, a division of The McGraw-Hill Companies, Inc.

Graph the integers on a number line. Then write them in order from least to greatest.

22 −3, −1, 5, 0, −5

<-+--+--+--+--+--+--+--+--+--+--+->
 −5 −4 −3 −2 −1 0 1 2 3 4 5

_____

23 8, −4, 5, −1, 9

<-+--+--+--+--+--+--+--+--+--+--+->
 −10 −8 −6 −4 −2 0 2 4 6 8 10

_____

24 10, −3, −2, 4, −6

<-+--+--+--+--+--+--+--+--+--+--+->
 −10 −8 −6 −4 −2 0 2 4 6 8 10

_____

25 −9, 8, −5, 2, 1

<-+--+--+--+--+--+--+--+--+--+--+->
 −10 −8 −6 −4 −2 0 2 4 6 8 10

_____

26 3, −4, −5, 7, −2

<-+--+--+--+--+--+--+--+--+--+--+->
 −10 −8 −6 −4 −2 0 2 4 6 8 10

_____

27 9, −4, −7, −9, 1

<-+--+--+--+--+--+--+--+--+--+--+->
 −10 −8 −6 −4 −2 0 2 4 6 8 10

_____

Write the integers from least to greatest.

28 64, −52, 18, −53, −16

_____

29 −55, 91, −102, 87, 78

_____

30 −47, 41, 74, 17, −71

_____

31 109, −901, 91, −19, 0

_____

Write the integers from greatest to least.

32 −256, −24, 182, −346, 265

_____

33 −802, 805, −806, −808, 809

_____

34 1,024; −565; 4,506; −5,656

_____

35 2,262; −262; −2,062; 6,262

_____

Copyright © Glencoe/McGraw-Hill, a division of The McGraw-Hill Companies, Inc.

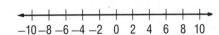

Solve.

36 **FOOTBALL** In a football game, players on the offensive team are not allowed to move before the ball is snapped. If this happens the team is given a five-yard penalty. What integer represents this penalty when a team breaks the rules?

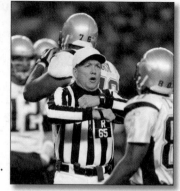

_____

37 **AWARDS** Carlos has worked for Louis Laboratories for 25 years. Last week, he received a service award for $250 dollars. What integer represents this award?

_____

38 **SCHOLARSHIPS** Last year, Investment in Education, Inc. gave a $1,500 scholarship to a student in Chicago, Illinois. What integer represents the scholarship that Investment in Education provided?

_____

39 **CHECKING ACCOUNTS** Mr. Fuentes deposited $457 into his checking account on Saturday. What integer represents the deposit?

_____

40 **WEATHER** The low temperature in Fairbanks, Alaska on February 1 was 13° below zero. What integer represents the temperature?

_____

Vocabulary Check **Write the vocabulary word that completes each sentence.**

41 _____ are the whole numbers and their opposites.

42 A(n) _____ number is a number greater than 0.

43 Numbers that are the same distance from 0 on a number line are

_____ integers.

44 **Writing in Math** Explain how to find the opposite of −6.

_____

_____

_____

STOP

Copyright © Glencoe/McGraw-Hill, a division of The McGraw-Hill Companies, Inc.

Add Integers

KEY Concept

The answer to an addition problem is called the **sum**.

To add positive numbers on a number line, begin at the *first* number. Move right the *same* number of spaces as the *second* number.

To add negative numbers on a number line, begin at the *first* number. Move left the same number of spaces as the second number.

Example: $-4 + (-2)$

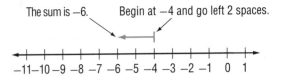

The sum is −6.   Begin at −4 and go left 2 spaces.

−11 −10 −9 −8 −7 −6 −5 −4 −3 −2 −1 0 1

The sum of an integer and its opposite is always zero. This is the **Inverse Property of Addition**.

$$17 + (-17) = 0 \text{ or } (-17) + 17 = 0$$

VOCABULARY

Commutative Property of Addition
the order in which two numbers are added does not change the sum
Example: $12 + 15 = 15 + 12$

Inverse Property of Addition
for any number, the sum of that number and its opposite is zero

The **Commutative Property of Addition** also applies when adding integers.

Example 1

What is the opposite of −6? Write an addition sentence to show the Inverse Property of Addition.

1. Graph the number. What number is the same distance from zero as −6? **6**

−10 −8 −6 −4 −2 0 2 4 6 8 10

2. Write an example of the inverse property using the two numbers.

$-6 + 6 = 0$

YOUR TURN!

What is the opposite of 4? Write an addition sentence to show the Inverse Property of Addition.

1. Graph the number. What number is the same distance from zero as 4? _____

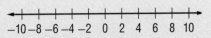

−10 −8 −6 −4 −2 0 2 4 6 8 10

2. Write an example of the inverse property using the two numbers.

$4 + \underline{\hspace{1cm}} = 0$

Copyright © Glencoe/McGraw-Hill, a division of The McGraw-Hill Companies, Inc.

Example 2

Find the sum of −10 and 7. Use algebra tiles.

1. Use 10 negative tiles to represent the first number.

2. Use 7 positive tiles to represent the second number.

3. A zero pair is made up of 1 positive tile and 1 negative tile. You can make 7 zero pairs.

4. There are 3 negative tiles left.

5. $-10 + 7 = -3$

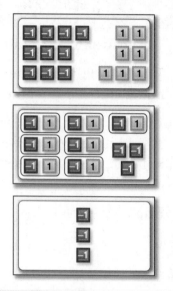

YOUR TURN!

Find the sum of −4 and 8. Use algebra tiles.

1. Use _____ to represent the first number.

2. Use _____ to represent the second number.

3. A zero pair is made up of 1 positive tile and 1 negative tile. You can make _____ zero pairs.

4. There are _____ left.

5. _____ + _____ = _____

Copyright © Glencoe/McGraw-Hill, a division of The McGraw-Hill Companies, Inc.

Copyright © Glencoe/McGraw-Hill, a division of The McGraw-Hill Companies, Inc.

Example 3

Find the sum of −8 and 6. Use a number line.

1. Graph the first number.

2. From the first number, go right on the number line.

3. You are at −2 on the number line.

4. Write the sum. **−8 + 6 = −2**

YOUR TURN!

Find the sum of −4 and 10. Use the number line.

1. Graph the first number.

2. From the first number, go _____ on the number line.

3. You are at _____ on the number line.

4. Write the sum. −4 + 10 = _____

Who is Correct?

Find the sum of 9 and −2.

Michael
9 + (−2) = −7

Sareeta
9 + (−2) = 7

Bianca
9 + (−2) = −11

Circle correct answer(s). Cross out incorrect answer(s).

▶ Guided Practice

What is the opposite of each number? Write an addition sentence to show the Inverse Property of Addition.

1 −3 _____

2 11 _____

3 25 _____

4 −62 _____

5 97 _____

6 201 _____

GO ON

Find each sum. Use algebra tiles.

7 $8 + (-1) =$ _____

8 $7 + (-4) =$ _____

9 $-7 + 9 =$ _____

10 $-3 + (-6) =$ _____

Step (by) Step **Practice**

11 Find the sum of 0 and -6. Use the number line.

    Step 1 Graph the first number.

    Step 2 From the first number, move
            _____ .

    Step 3 Where are you on the number line?
            _____

    Step 4 Write the sum. _____

Find each sum. Use the number line.

12 $-11 + 4 =$ _____   Start at _____ and move _____ _____ places.

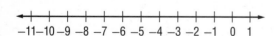

13 $-3 + (-6) =$ _____

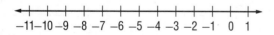

14 $-2 + (-3) =$ _____

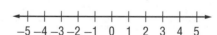

Copyright © Glencoe/McGraw-Hill, a division of The McGraw-Hill Companies, Inc.

Find each sum.

15 $-4 + (-5) =$ _____

16 $10 + (-3) =$ _____

17 $-7 + (-1) =$ _____

18 $15 + (-5) =$ _____

19 $-17 + (-11) =$ _____

20 $25 + (-15) =$ _____

Step by Step Problem-Solving Practice

Solve.

Problem-Solving Strategies
☑ Draw a picture.
☐ Use logical reasoning.
☐ Make a table.
☐ Solve a simpler problem.
☐ Work backward.

21 **ELEVATION** Osceola County has the highest point of elevation in the state of Iowa. This point is approximately 500 meters above sea level. One point in Lee County, Iowa has the lowest elevation in the state. There is approximately 350 meters difference between the two points. What is the elevation of the lowest point in Iowa?

Understand Read the problem. Write what you know.

The highest point is at an elevation of _____.

The difference between the points is _____.

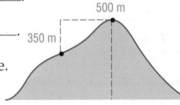

500 m
350 m

Plan Pick a strategy. One strategy is to draw a picture.

Solve Begin at 500 meters and go down 350 meters.

_____

Check Does the answer make sense? Look over your solution. Did you answer the question?

22 **VENDING MACHINES** Hanako puts 12 quarters into a vending machine. She uses $2.25 of her credit. How much credit does she have left?

Check off each step.

_____ Understand: I underlined key words.

_____ Plan: To solve the problem I will _____.

_____ Solve: The answer is _____.

_____ Check: I checked my answer by _____.

Copyright © Glencoe/McGraw-Hill, a division of The McGraw-Hill Companies, Inc.

GO ON

23 **STOCKS** Reynaldo has stock in the Capital Calendar Company. At the beginning of the day the price per share was $42.50. At the end of the day, the price per share had gone down by $5.00. What was the price per share at the end of the day?

_____

24 **Reflect** Write and solve two addition sentences using positive and negative integers to show the Inverse Property of Addition. Explain your answer.

_____

_____

▶ Skills, Concepts, and Problem Solving

What is the opposite of each number? Write an addition sentence to show the Inverse Property of Addition.

25 −9 _____

26 16 _____

27 59 _____

28 −73 _____

Find each sum. Use algebra tiles.

29 3 + (−6) = _____

30 −4 + 6 = _____

Find each sum. Use the number line.

31 −2 + 3 = _____

−5 −4 −3 −2 −1 0 1 2 3 4 5

32 3 + (−3) = _____

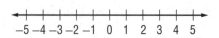
−5 −4 −3 −2 −1 0 1 2 3 4 5

33 4 + (−10) = _____

−10 −8 −6 −4 −2 0 2 4 6 8 10

34 −6 + (−2) = _____

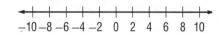
−10 −8 −6 −4 −2 0 2 4 6 8 10

Copyright © Glencoe/McGraw-Hill, a division of The McGraw-Hill Companies, Inc.

Find each sum.

35 $28 + (-12) = $ _____

36 $-46 + (-35) = $ _____

37 $-42 + 41 = $ _____

38 $-37 + (-7) = $ _____

39 $-52 + 12 = $ _____

40 $-105 + 80 = $ _____

41 **TRAVEL** An airplane has 66 passengers traveling from Columbus, Ohio to Atlanta, Georgia. There are 6 first-class passengers. The first-class passengers exit the airplane first. How many people remain on the airplane?

_____

42 **SPORTING EVENTS** Christopher is attending a basketball game. He walked up to row 32 to buy a bag of peanuts for his sister. Christopher walked down 15 rows to arrive at his seat. In what row is Christopher's seat?

_____

Vocabulary Check **Write the vocabulary word that completes the sentence.**

43 The property that states that the order in which numbers are added does not affect the sum is the _____ .

44 **Writing in Math** Explain how to use the number line to find the sum of -7 and -3.

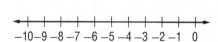

_____

 Spiral Review

Write the integers from least to greatest. (Lesson 1-1, p. 4)

45 $-18, -10, 20, 14, -13$ _____

46 $68, -42, 91, -19, 35$ _____

47 **FINANCES** You spend $37. What integer represents your money? _____

STOP

Graph the integers on a number line. Then write them in order from least to greatest.

1 −13, −2, 3, −3 _____

2 −2, 6, 2, −5, 0 _____

Write <, =, or > in each circle to make a true statement.

3 −6 ◯ 0

4 10 ◯ −15

5 2 ◯ −12

What is the opposite of each number? Use it to show the Inverse Property of Addition.

6 3 _____

7 −18 _____

Find each sum. Use algebra tiles.

8 −5 + (−3) = _____

9 6 + (−9) = _____

Find each sum. Use the number line.

10 −6 + (−5) = _____

11 −7 + 9 = _____

Solve.

12 **CALORIES** Every 10 steps burns 5 Calories. Every taco adds 185 Calories. If Beth ate 1 taco and took 100 steps, how many calories would she have gained or lost?

_____

13 **FOOTBALL** Matthew's team was at their own 35-yard line during a football game. They lost 15 yards. What yard line were they on for the next play?

_____

Copyright © Glencoe/McGraw-Hill, a division of The McGraw-Hill Companies, Inc.

Copyright © Glencoe/McGraw-Hill, a division of The McGraw-Hill Companies, Inc.

Subtract Integers

KEY Concept

Subtraction can be defined as adding the **opposite** of a number.

$3 - 7$ can be written as the addition expression $3 + (-7)$.

You can use the number line to show the sum.

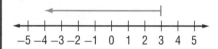

Start at the first number.
Since the next number is negative,
move to the *left* 7 places. The sum is –4.

The **absolute value** of a number is the distance the number is from zero. The symbol for absolute value of the number x is $|x|$.

$|-4| = 4$ and $|4| = 4$

Finding the absolute value of numbers can help you when adding and subtracting integers.

Subtracting Integers			
Subtraction	Rewritten as Addition	Signs	Answer
$-2 - 6$	$-2 + (-6)$	same	-8
$2 - 6$	$2 + (-6)$	different	-4
$-2 - (-6)$	$-2 + 6$	different	4
$2 - (-6)$	$2 + 6$	same	8

VOCABULARY

absolute value
the distance between a number and 0 on a number line

opposites
numbers that are the same distance from zero in opposite directions

Rewrite subtraction problems as adding the opposite. The rules given above for determining the sign of an answer are used for both addition and subtraction problems.

GO ON

Example 1

Find the difference of −4 and −2. Use algebra tiles.

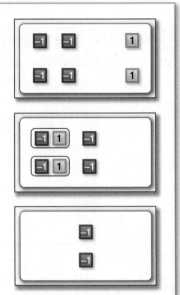

1. Write the subtraction expression. −4 − (−2)

2. Add the opposite.
   Write the addition expression. −4 + 2

3. Use 4 negative tiles and 2 positive tiles to represent the numbers.

4. You can make 2 zero pairs.

5. There are 2 negative tiles left.

6. Write the difference. −4 − (−2) = −2

YOUR TURN!

Find the difference of −5 and −8. Use algebra tiles.

1. Write the subtraction expression. _____

2. Add the opposite.
   Write the addition expression. _____

3. Use _____ and _____ to represent the numbers.

4. You can make _____ zero pairs.

5. There are _____ left.

6. Write the difference. _____

Copyright © Glencoe/McGraw-Hill, a division of The McGraw-Hill Companies, Inc.

Example 2

Find the difference of −3 and −5. Use the number line.

1. Write the subtraction expression.

   −3 − (−5)

2. To subtract integers, add the opposite.
   Write the addition expression.

   −3 + 5

3. Graph the first number.

4. The sign of the second integer is positive.
   Move right on the number line.

5. You are at 2 on the number line.

6. Write the difference. −3 − (−5) = 2

YOUR TURN!

Find the difference of 1 and 4. Use the number line.

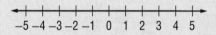

1. Write the subtraction expression.

   _____

2. To subtract integers, add the opposite.
   Write the addition expression.

   _____

3. Graph the first number.

4. The sign of the second integer is _____.
   Move _____ on the number line.

5. You are at _____ on the number line.

6. Write the difference. _____

GO ON

Copyright © Glencoe/McGraw-Hill, a division of The McGraw-Hill Companies, Inc.

Example 3

Does −5 or 3 have a greater absolute value?

1. −5 is 5 units from 0. So, |−5| = 5.

2. 3 is 3 units from 0. So, |3| = 3.

3. Which integer has the greater absolute value?

   −5

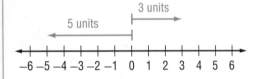

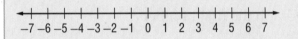

YOUR TURN!

Does 5 or −6 have a greater absolute value?

1. How far is 5 from 0? _____

   So, |5| = _____.

2. How far is −6 from 0? _____

   So, |−6| = _____.

3. Which integer has the greater absolute value?

   _____

Who is Correct?

Find the difference of −3 and 5. Use algebra tiles.

Circle correct answer(s). Cross out incorrect answer(s).

Copyright © Glencoe/McGraw-Hill, a division of The McGraw-Hill Companies, Inc.

Guided Practice

Find each difference. Use algebra tiles.

1 $4 - 2 = $ _____

2 $-6 - (-4) = $ _____

Step by Step Practice

3 Find the difference of -3 and -6. Use the number line.

Step 1 Write the subtraction expression. _____

Step 2 To subtract integers, add the opposite.
Write the addition expression. _____
This is the new expression.

Step 3 Graph the first number.

Step 4 The sign of the second integer is _____.
Which direction will you go on the
number line? _____

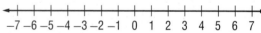

Step 5 Where are you on the number
line? _____

Step 6 Write the difference. _____

Find each difference. Use the number line.

4 $3 - (-5)$ addition sentence: _____ + _____
sum: _____ + _____ = _____

5 $4 - (-9) = $ _____

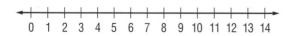

6 $-5 - (-7) = $ _____

GO ON

Copyright © Glencoe/McGraw-Hill, a division of The McGraw-Hill Companies, Inc.

Find each difference. Use algebra tiles.

7 $-12 - 7 =$ _____

8 $-20 - 11 =$ _____

9 $15 - (-16) =$ _____

10 $-23 - (-21) =$ _____

Which number has the greater absolute value?

11 3 or -5 _____

12 15 or -15 _____

13 10 or -7 _____

Step by Step Problem-Solving Practice

Problem-Solving Strategies
- ☐ Draw a diagram.
- ☐ Use logical reasoning.
- ☑ Make a table.
- ☐ Solve a simpler problem.
- ☐ Act it out.

Solve.

14 **GAS PRICES** Josephina used estimates to compare the cost of 10 gallons of gas. In February 2006, the cost was $22. During 2007, the cost was about $1 lower. In February 2008, the cost was $8 higher than 2007. What was the cost of 10 gallons of gas in February 2008?

Understand Read the problem. Write what you know.

Ten gallons of gas cost about _____ in 2006.

Ten gallons of gas cost _____ less in 2007.

The price rose by _____ in 2008.

Plan Pick a strategy. One strategy is to make a table.

Solve Follow the changes in gas prices by using the table.

_____ − _____ + _____ = _____

Change ($)	Cost
	$22
−1	
+8	

Check Use algebra tiles to check your answer.

15 **SPORTS** Dion took a vacation in North Carolina. He went scuba diving to a depth of 240 feet below sea level . Later that week, he hiked to the peak of Mount Mitchell at 6,700 feet above sea level. What is the difference in altitudes between the two locations?

Check off each step.

_____ **Understand: I underlined key words.**

_____ **Plan: To solve the problem, I will** _____.

_____ **Solve: The answer is** _____.

_____ **Check: I checked my answer by** _____.

Copyright © Glencoe/McGraw-Hill, a division of The McGraw-Hill Companies, Inc.

16 **WEATHER** Evelina was comparing the record temperatures for January in Fairbanks, Alaska. The highest recorded temperature was 50°F in 1981. The lowest recorded temperature was −66°F in 1934. What is the difference between these two records?

_____

17 **Reflect** Explain how to determine the absolute value of −12.

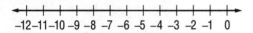

_____

_____

Skills, Concepts, and Problem Solving

Find each difference. Use algebra tiles.

18 $2 - 6 =$ _____

19 $-3 - 4 =$ _____

Find each difference. Use the number line.

20 $-4 - 7 =$ _____

21 $-9 - (-10) =$ _____

Which number has the greater absolute value?

22 −5 or 5 _____

23 −2 or −12 _____

24 5 or −3 _____

Find each difference.

25 $-4 - (-2) =$ _____

26 $-16 - 60 =$ _____

27 $-5 - 48 =$ _____

28 $-96 - (-56) =$ _____

GO ON

29 **NUTRITION**   For breakfast, Grace ate a bagel and cream cheese. Her breakfast had 390 Calories. Later that morning Grace ran for 45 minutes and burned 460 Calories. What is her net gain of Calories?

_____

30 **SPORTS**   Brandon went to watch his little brother's football game on Saturday. During one play, his brother ran the wrong direction for 7 yards. He later turned around and ran in the correct direction for 42 yards. What is the net gain of yards?

_____

Vocabulary Check   **Write the vocabulary word that completes the sentence.**

31 The _____ of a number is the distance the number is from zero.

32 **Writing in Math**   Sadzi lives 400 miles east of Albuquerque, New Mexico. Her friend Khadijah lives 1,750 miles west of Albuquerque, New Mexico. What is the difference in distance between the friends' homes? Explain how to find the distance using a number line.

_____

_____

_____

▶ **Spiral Review**

What is the opposite of each number? Use it to show the Inverse Property of Addition. (Lesson 1–2, p. 11)

33 8 _____   **34** −19 _____

35 −47 _____   **36** 241 _____

37 **FINANCES**   Emma borrowed $15 from her sister last Monday. This Monday, Emma borrowed another $17. How much money does Emma have? Use integers to write an equation.

_____

Copyright © Glencoe/McGraw-Hill, a division of The McGraw-Hill Companies, Inc.

Multiply Integers

KEY Concept

To find the **product** of two integers, multiply the absolute values of the **factors** and then determine the correct sign of the answer.

If the signs are the same, then the sign of the product is positive.

$$4 \cdot 4 = 16 \text{ and } -4 \cdot (-4) = 16$$

If the signs are different, then the sign of the product is negative.

$$3 \cdot (-2) = -6 \text{ and } -3 \cdot 2 = -6$$

Multiplication of integers can be shown on a number line as repeated addition.

$$-2 \cdot 3 \text{ is } (-2) + (-2) + (-2) \text{ or } -6$$

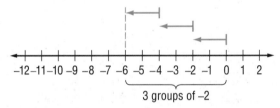

3 groups of –2

The multiplication properties apply to integers.

Properties of Multiplication		
Property	**States that...**	**Example**
Zero	any number times zero equals zero.	$-3 \cdot 0 = 0$
Identity	any number times 1 equals that number.	$-2 \cdot 1 = -2$
Commutative	the order in which numbers are multiplied does not change the product.	$-4 \cdot 2 = 2 \cdot (-4)$
Associative	the manner in which factors are grouped does not change the product.	$3 \cdot (-4 \cdot 4) = 4 \cdot [(-4) \cdot 3]$

VOCABULARY

Associative Properties
the way in which numbers are grouped does not change the sum or product

Commutative Properties
the order in which numbers are added or multiplied does not change the sum or product

Distributive Property
to multiply a sum by a number, multiply each addend by the number outside the parentheses

factor
a number that divides into a whole number with a remainder of zero; also a number that is multiplied by another number

product
the answer or result to a multiplication problem; it also refers to expressing a number as the product of its factors

You can use the Distributive Property to simplify addition and multiplication problems.

$$8(10 + (-2)) = (8 \cdot 10) + (8 \cdot (-2))$$

Copyright © Glencoe/McGraw-Hill, a division of The McGraw-Hill Companies, Inc.

Example 1

Find 2 · (−4). Use a number line.

1. Identify the first number in the expression. **2**
   This is the number of times the group is repeated.

2. Identify the second number in the expression. **−4**
   This is the group size.

3. Draw a number line. Mark off **2** groups of **−4**.

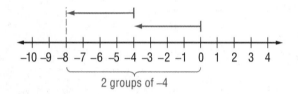

−10 −9 −8 −7 −6 −5 −4 −3 −2 −1 0 1 2 3 4

2 groups of −4

4. The signs are different, so the product is negative.

5. Write the product.
   −8

YOUR TURN!

Find 3 · (−2). Use a number line.

1. What is the first number in the expression? _____
   This is the number of times the group is repeated.

2. Identify the second number in the expression. _____
   This is the group size.

3. Draw a number line. Mark off _____ groups of _____.

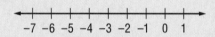

−7 −6 −5 −4 −3 −2 −1 0 1

4. Are the signs the same or different? _____ Will the product be positive or negative? _____

5. Write the product.
   _____

Example 2

Find 6 · (−3) by multiplying absolute values.

1. Find the absolute value of each.
   $|6| = 6$ and $|-3| = 3$

2. Multiply the absolute values of the numbers.
   $6 · 3 = 18$

3. The signs are different, so the product is negative.

4. Write the product with the sign.
   −18

YOUR TURN!

Find 7 · (−5) by multiplying absolute values.

1. Find the absolute value of each.
   _____ and _____

2. Multiply the absolute values of the numbers.
   _____ · _____ = _____

3. Are the signs the same or different? _____ Will the product be positive or negative? _____

4. Write the product with the sign.
   _____

Copyright © Glencoe/McGraw-Hill, a division of The McGraw-Hill Companies, Inc.

Example 3

Simplify. Name the multiplication properties that are used in each step.

$(3 \cdot (-8)) \cdot 1 = 3 \cdot (-8 \cdot 1)$ Associative Property

$= 3 \cdot (-8)$ Identity Property

$= -24$

YOUR TURN!

Simplify. Name the multiplication properties that are used in each step.

$(-12 \cdot 0) = (0 \cdot (-12))$ _____

$= 0$ _____

Who is Correct?

Find $7 \cdot (-10)$.

Alejandro
$7 \cdot (-10) = -3$

Heather
$7 \cdot (-10) = -70$

Kendrick
$7 \cdot (-10) = 70$

Circle correct answer(s). Cross out incorrect answer(s).

 Guided Practice

Find each product. Use a number line.

1 $-1 \cdot 7 =$ _____

2 $4 \cdot (-3) =$ _____

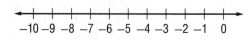

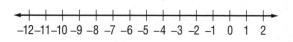

Copyright © Glencoe/McGraw-Hill, a division of The McGraw-Hill Companies, Inc.

GO ON

Copyright © Glencoe/McGraw-Hill, a division of The McGraw-Hill Companies, Inc.

Step by Step Practice

3 Find $-8 \cdot (-7)$.

Step 1 Find the absolute value of each. _____

Step 2 Multiply the absolute values of the numbers.
_____ • _____ = _____

Step 3 Are the signs the same or different? _____

Step 4 Will the product be positive or negative? _____

Step 5 Write the product with the sign. _____

Step 6 Check the sign to make sure it is correct.
$(-) \cdot (-) = (+)$

_____

Find each product.

4 $-8 \cdot (-6)$ absolute value: _____ • _____ = _____
product: _____ sign: _____

5 $7 \cdot (-9) = $ _____ **6** $5 \cdot 7 = $ _____ **7** $-4 \cdot 8 = $ _____ **8** $-9 \cdot (-6) = $ _____

Step by Step Problem-Solving Practice

Solve.

9 **EVAPORATION** The height of the water in Fernando's swimming pool decreases 2 centimeters per week due to evaporation. What is the change in the height of the water over a six-week period, due to evaporation?

Problem-Solving Strategies
☑ Draw a diagram.
☐ Use logical reasoning.
☐ Guess and check.
☐ Solve a simpler problem.
☐ Work backwards.

Understand	Read the problem. Write what you know. The height of the pool water changed by _____ cm each week for _____ weeks.
Plan	Pick a strategy. One strategy is to draw a number line to represent the height of the pool water.
Solve	Use the number line to find the change in value. Draw a number line. Mark off _____ groups of _____. The change in the height of the water is _____ cm.
Check	Use the Commutative Property to multiply the factors in a different order. Your product should be the same.

10 HIKING   Hayden and his friend went hiking in the Rocky Mountains. They descended the mountain at a rate of 7 feet per minute. What is the change in elevation of the hikers after 15 minutes?

Check off each step.

_____ **Understand: I underlined key words.**

_____ **Plan: To solve the problem I will** _____.

_____ **Solve: The answer is** _____.

_____ **Check: I checked my answer by** _____.

11 WEATHER   During a certain week, the temperature in Pensacola, Florida dropped 3°F each day. What is the change in temperature in Pensacola over 5 days?

_____

12 Reflect   Explain how to use a number line to multiply $3 \cdot (-4)$.

_____

_____

▶ Skills, Concepts, and Problem Solving

Find each product. Use a number line.

13 $2 \cdot (-5) =$ _____

14 $3 \cdot (-6) =$ _____

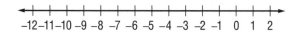

-12 -11 -10 -9 -8 -7 -6 -5 -4 -3 -2 -1 0 1 2

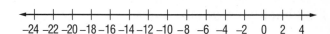

-24 -22 -20 -18 -16 -14 -12 -10 -8 -6 -4 -2 0 2 4

Find each product.

15 $3 \cdot (-4) =$ _____

16 $-9 \cdot 2 =$ _____

17 $-8 \cdot (-10) =$ _____

18 $4 \cdot (-7) =$ _____

19 $6 \cdot (-8) =$ _____

20 $-12 \cdot (-5) =$ _____

Find the missing number. Name the multiplication property.

21 $-84 \cdot$ _____ $= -84$ _____

22 $(4 \cdot 7) \cdot (-8) = 4 \cdot [$ _____ $\cdot (-8)]$ _____

Copyright © Glencoe/McGraw-Hill, a division of The McGraw-Hill Companies, Inc.

Solve.

23 **POPULATION** The population of Jefferson Middle School has decreased by 15 students every year for the past three years. What is the change in the population?

_____

24 **FINANCES** Jasmine bought a stock for $27 a share. The stock lost $3 for each of the next 5 months. How much did each share of stock lose in value?

_____

Vocabulary Check **Write the vocabulary word that completes each sentence.**

25 The _____ states that when you multiply a number by 0, the product is zero.

26 In the equation $-3 \cdot (-7) = 21$, the integers _____ and _____ are factors.

27 **Writing in Math** Audrey worked the following problem. What mistake did she make?

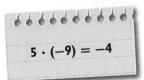

$5 \cdot (-9) = -4$

_____

_____

_____

▶ Spiral Review

Find each sum or difference. (Lesson 1-2, p. 11 and Lesson 1-3, p. 19)

28 $5 + (-7) =$ _____

29 $-110 - (-60) =$ _____

30 $-44 - (-32) =$ _____

31 $-81 + 100 =$ _____

Solve. (Lesson 1-1, p. 4)

32 **AIRPLANES** An airplane was cruising at an altitude of 27,000 feet. The pilot descended 350 feet to avoid a thunderstorm. What integer describes the airplane's altitude?

_____

STOP

Copyright © Glencoe/McGraw-Hill, a division of The McGraw-Hill Companies, Inc.

Copyright © Glencoe/McGraw-Hill, a division of The McGraw-Hill Companies, Inc.

Lesson 1-5 Divide Integers

KEY Concept

Division is the inverse operation for multiplication. You use the multiplication facts whenever you divide integers.

If the signs are the same, then the **quotient** is positive.

$72 \div 8 = 9$
$(-72) \div (-8) = 9$

If the signs are different, then the quotient is negative.

$(-72) \div 8 = -9$
$72 \div (-8) = -9$

VOCABULARY

dividend
the number that is being divided

$$\text{divisor} \rightarrow 4\overline{)8} \begin{matrix} 2 \leftarrow \text{quotient} \\ \leftarrow \text{dividend} \end{matrix}$$

divisor
the number by which the dividend is being divided

quotient
the answer or result of a division problem

You can find the quotient of two integers by using a related multiplication sentence, $-8 \cdot (-9) = 72$. So, $72 \div (-8) = -9$.

Example 1

Find $48 \div (-6)$.

1. The signs are different. The quotient will be negative.
2. Find the absolute value of each. $|48| = 48$ and $|-6| = 6$
3. Divide the absolute values of the numbers. $48 \div 6 = 8$
4. Write the quotient with a negative sign. -8

YOUR TURN!

Find $-88 \div 8$.

1. Are the signs the same or different? _____
2. Find the absolute value of each. _____
3. Divide the absolute values of the numbers. _____ ÷ _____ = _____
4. Write the quotient with the sign. _____

GO ON

Example 2

Find $\dfrac{-42}{-6}$.

1. The signs are the same.

2. The quotient will be positive.

3. Divide the absolute values of the numbers.
   $42 \div 6 = 7$

4. Write the quotient. 7

YOUR TURN!

Find $\dfrac{-35}{-5}$.

1. Are the signs the same or different?
   _____

2. Will the quotient be positive or negative?
   _____

3. Divide the absolute values of the numbers.

   _____ ÷ _____ = _____

4. Write the quotient. _____

Who is Correct?

Simplify $-63 \div (-9)$.

Grayson
$-63 \div (-9) = -72$

Pilar
$-63 \div (-9) = -7$

Terrell
$-63 \div (-9) = 7$

Circle correct answer(s). Cross out incorrect answer(s).

 Guided Practice

Find each quotient.

1 $12 \div 1 = $ _____

2 $-42 \div 1 = $ _____

3 $-27 \div (-1) = $ _____

4 $93 \div (-1) = $ _____

5 $72 \div (-12) = $ _____

6 $-64 \div (-4) = $ _____

7 $-60 \div 5 = $ _____

8 $44 \div 2 = $ _____

Copyright © Glencoe/McGraw-Hill, a division of The McGraw-Hill Companies, Inc.

Step by Step Practice

9 Find $-12 \div 2$.

Step 1 Are the signs the same or different? _____

Step 2 Will the quotient be positive or negative? _____

Step 3 Divide the absolute values of the numbers.

_____

Step 4 Write the quotient. _____

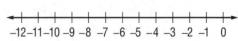

Check the number line. The integer -12 can be divided into 2 groups of -6.

Find each quotient.

10 $-9 \div 3$

signs: _____ ÷ _____ = _____

quotient: _____

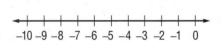

11 $\dfrac{-8}{4}$

signs: _____ ÷ _____ = _____

quotient: _____

12 $-36 \div 3 =$ _____

13 $52 \div (-13) =$ _____

14 $-20 \div (-4) =$ _____

15 $-50 \div (-5) =$ _____

16 $\dfrac{-14}{2} =$ _____

17 $\dfrac{-23}{1} =$ _____

18 $\dfrac{56}{-4} =$ _____

19 $\dfrac{-36}{12} =$ _____

20 $\dfrac{-200}{-25} =$ _____

21 $\dfrac{-72}{-12} =$ _____

GO ON

Copyright © Glencoe/McGraw-Hill, a division of The McGraw-Hill Companies, Inc.

Step by Step Problem-Solving Practice

Solve.

Problem-Solving Strategies

☑ Use a model.
☐ Use logical reasoning.
☐ Make a table.
☐ Guess and check.
☐ Solve a simpler problem.

22 FINANCES Mr. Sullivan withdrew $720 from his savings account over the last 6 months. If he withdrew the same amount each month, what was the amount of each monthly withdrawal?

Understand Read the problem. Write what you know.

Mr. Sullivan withdrew a total of _____ over a period of _____ months.

Plan Pick a strategy. One strategy is to use a model.

Draw a fraction bar worth a total of $720. Divide it into 6 equal parts. How much is each part worth?

Solve Divide. _____ ÷ _____ = _____
Mr. Sullivan withdrew _____ per month.

Check Multiply to check your answer.

_____ • 6 = _____

23 EROSION One shoreline along the Gulf of Mexico has receded by 42 feet over the past 7 years. If the beach has eroded at the same rate each year, what is the average yearly rate of erosion?

Check off each step.

_____ Understand: I underlined key words.

_____ Plan: To solve the problem I will _____.

_____ Solve: The answer is _____

_____

_____ Check: I checked my answer by _____.

Copyright © Glencoe/McGraw-Hill, a division of The McGraw-Hill Companies, Inc.

GO ON

24 **AVIATION** A plane descends from an altitude of 7,500 feet at a rate of 500 feet per minute. How long does it take for the plane to land?

_____

25 **SWIMMING POOLS** Mrs. Whitmore is draining the family pool for the winter. The pool holds 540 gallons of water. If the pool drains at a rate of 90 gallons per hour, how many hours will it be before the pool is empty?

_____

26 **Reflect** Lorenzo completed the following equation. $-27 \div 3 = -9$
He plans to check his answer using the following
multiplication expression. $-3 \cdot (-9)$
Is Lorenzo's plan correct? Explain your answer.

_____

_____

 ## Skills, Concepts, and Problem Solving

Find each quotient.

27 $-10 \div 5 =$ _____

signs: _____ $\div$ _____ = _____

quotient: _____

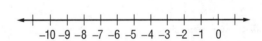

28 $\dfrac{-14}{7} =$ _____

signs: _____ $\div$ _____ = _____

quotient: _____

29 $-64 \div 8 =$ _____

30 $54 \div (-9) =$ _____

31 $-72 \div (-6) =$ _____

32 $70 \div 14 =$ _____

33 $\dfrac{-49}{7} =$ _____

34 $\dfrac{42}{-6} =$ _____

35 $\dfrac{-54}{-3} =$ _____

36 $\dfrac{36}{6} =$ _____

GO ON

Copyright © Glencoe/McGraw-Hill, a division of The McGraw-Hill Companies, Inc.

Solve.

37 **INVESTMENTS**   Last year, Yolanda had a total of $675 deducted from her paycheck and sent into her individual retirement account (IRA). She made 3 payments to her individual retirement account. What integer represents the amount of money that was deducted for each IRA payment?

_____

38 **WEATHER**   The temperature at the base of the mountain is 37°F. At the top of the mountain, which is 5,000 feet high, the temperature is −13°F. Khalid is driving up to the peak of the mountain. What integer describes the change in temperature for every 200 feet he travels?

_____

_____

Vocabulary Check   **Write the vocabulary word that completes each sentence.**

39   The _____ is the result of a division problem.

40   A(n) _____ number is less than zero.

41   **Writing in Math**   Celeste worked the problem on the right. What mistake did she make?

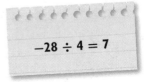

$$-28 \div 4 = 7$$

_____

_____

 Spiral Review

Find the missing number. Name the multiplication property. (Lesson 1–4, p. 27)

42 $-84 \cdot$ _____ $= 0$ _____

43 $4 \cdot (-7) =$ _____ $\cdot 4$ _____

Solve. (Lesson 1–3, p. 19)

44   **NUTRITION**   Elena ate a cereal and fruit bar and a glass of orange juice for breakfast. Her breakfast had 240 Calories. Later that morning Elena bicycled for 45 minutes and burned 354 Calories. What is her net gain of Calories? _____

Copyright © Glencoe/McGraw-Hill, a division of The McGraw-Hill Companies, Inc.

Find each difference. Use algebra tiles.

1 $-8 - (-2) =$ _____

2 $-3 - 2 =$ _____

Which number has the greater absolute value?

3 -16 or 3 _____

4 -12 or 12 _____

5 4 or 7 _____

6 -8 or -6 _____

Find each product.

7 $7 \cdot (-4) =$ _____

8 $-9 \cdot 5 =$ _____

9 $-9 \cdot (-12) =$ _____

10 $8 \cdot (-7) =$ _____

11 $13 \cdot (-8) =$ _____

12 $-14 \cdot (-5) =$ _____

Find each quotient.

13 $-36 \div (-4) =$ _____

14 $-55 \div 5 =$ _____

15 $-72 \div (-12) =$ _____

16 $\dfrac{56}{-8} =$ _____

17 $\dfrac{-64}{4} =$ _____

18 $\dfrac{-120}{-12} =$ _____

Solve.

19 FINANCES Mr. Wilkerson bought stock for $27 a share. Each share lost $3 for each of the next 4 months. How much has each share lost in value?

_____

20 TRAVEL Casandra hiked The Big Tree Trail in the Oregon Caves National Monument. She hiked up the trail 1,100 feet. Then she began her descent back to the trailhead. So far, she has descended 550 feet. What integer describes her elevation?

_____

Copyright © Glencoe/McGraw-Hill, a division of The McGraw-Hill Companies, Inc.

Vocabulary and Concept Check

absolute value, *p. 19*

Commutative Property of Addition, *p. 11*

dividend, *p. 33*

divisor, *p. 33*

Identity Property of Multiplication, *p. 27*

integers, *p. 4*

Inverse Property of Addition, *p. 11*

negative number, *p. 4*

opposite, *p. 4*

positive number, *p. 4*

quotient, *p. 33*

whole number, *p. 4*

Zero Property of Multiplication, *p. 27*

Write the vocabulary word that completes each sentence. Not all vocabulary terms will be used.

1 0, 1, 2, 3, 4 … are _____.

2 The _____ is the distance between a number and 0 on a number line.

3 A number less than zero is a(n) _____.

4 …−3, −2, −1, 0, 1, 2, 3 … are _____.

5 Two different numbers that are the same distance from 0 on a number line are _____ numbers.

Write the name of the property shown below.

6 −5 · 0 = 0 _____

7 9 + (−9) = 0 _____

8 75 · 1 = 75 _____

Copyright © Glencoe/McGraw-Hill, a division of The McGraw-Hill Companies, Inc.

Lesson Review

1-1 Model Integers (pp. 4–10)

Write <, =, or > in each circle to make a true statement.

9 2 $\bigcirc$ −3

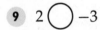

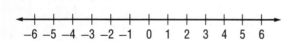

10 −6 $\bigcirc$ 6

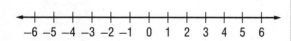

Example 1

Use <, =, or > to compare −5 and 3.

1. Graph both numbers on the number line.

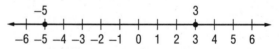

2. The number farthest to the right is 3, so it is the greater number.

3. Since −5 is less than 3, you need to use the "less than" symbol.

   −5 < 3

1-2 Add Integers (pp. 11–17)

Find each sum. Use algebra tiles.

11 $4 + (-3) =$ _____

12 $-5 + (-2) =$ _____

13 $-7 + 5 =$ _____

Find each sum.

14 $-1 + 8 =$ _____

15 $8 + (-7) =$ _____

16 $-7 + (-5) =$ _____

17 $15 + (-4) =$ _____

18 $9 + 23 =$ _____

19 $-27 + (-10) =$ _____

Example 2

Find the sum of −6 and 2. Use algebra tiles.

1. Use six negative tiles and two positive tiles to represent the numbers.

2. You can make 2 zero pairs.

3. There are four negative tiles left.

4. The sum is 4.
$$-6 + 2 = 4$$

Example 3

Find the sum of 5 and −6. Use the number line.

1. Graph the first number.

2. From the first number, go left on the number line.

3. You are at −1 on the number line.

4. Write the sum.
$$5 + (-6) = -1$$

Copyright © Glencoe/McGraw-Hill, a division of The McGraw-Hill Companies, Inc.

1-3 Subtract Integers (pp. 19–26)

Find each difference. Use algebra tiles.

20 $4 - (-3) = $ _____

21 $-6 - (-2) = $ _____

22 $-4 - 5 = $ _____

Find each difference.

23 $-5 - 8 = $ _____

24 $14 - (-3) = $ _____

25 $6 - 18 = $ _____

26 $-3 - (-8) = $ _____

Which number has the greater absolute value?

27 2 or −9 _____

28 −15 or −3 _____

Example 4

Find the difference of −5 and −3. Use algebra tiles.

1. Write the subtraction expression. $-5 - (-3)$

2. Write the addition expression. $-5 + 3$

3. Use five negative tiles and three positive tiles to represent the numbers.

4. You can make 3 zero pairs.

5. There are 2 negative tiles left.

6. The difference is −2.
$$-5 - (-3) = -2$$

Example 5

Which number has the greater absolute value?

$|5|$ or $|-3|$

1. 5 is 5 units from 0.

2. −3 is 3 units from 0.

3. Which integer has the greater absolute value? **5**

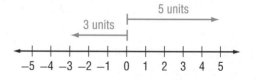

Copyright © Glencoe/McGraw-Hill, a division of The McGraw-Hill Companies, Inc.

1-4 Multiply Integers (pp. 27–32)

Find each product.

29 −8 · (−3)

absolute value: _____ · _____ = _____

sign: _____   product: _____

30 −13 · (5)

absolute value: _____ · _____ = _____

sign: _____   product: _____

31 8 · (−6) =

_____

32 −9 · (−12) =

_____

33 −6 · (−15) =

_____

Example 6

Find −7 · (−13) by multiplying absolute values.

1. Find the absolute value of each.
   $|-7| = 7$ and $|-13| = 13$

2. Multiply the absolute values of the numbers.
   $7 \cdot 13 = 91$

3. The signs are the same. The product is positive.

4. Write the product with the sign.
   91

1-5 Divide Integers (pp. 33–38)

Find each quotient.

34 $\dfrac{-48}{-6} =$

_____

35 −49 ÷ 7 =

_____

36 $\dfrac{84}{-12} =$

_____

37 −99 ÷ (−11) =

_____

Example 7

Find 36 ÷ (−3).

1. Find the absolute value of each.
   $|36| = 36$ and $|-3| = 3$

2. The signs are different.
   The quotient will be negative.

3. Divide the absolute values of the numbers.
   $36 \div 3 = 12$

4. Write the quotient with a negative sign.
   −12

Copyright © Glencoe/McGraw-Hill, a division of The McGraw-Hill Companies, Inc.

Chapter Test

1 Graph $-8, 3, 0, -5, 7$. Then write the numbers in order from least to greatest.

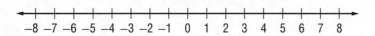

_____

Write <, =, or > in each circle to make a true statement.

2 $-7 \bigcirc -10$

3 $-12 \bigcirc 2$

4 $0 \bigcirc -17$

5 $|-3| \bigcirc |3|$

Find each sum.

6 $-5 + (-4) =$ _____

7 $(-3) + 12 =$ _____

8 What is the opposite of 18? Use it to show the Inverse Property of Addition.

_____

Find each difference.

9 $-2 - (-4) =$ _____

10 $-7 - (-1) =$ _____

11 $7 - (-2) =$ _____

12 $6 - 9 =$ _____

Which number has the greater absolute value?

13 -12 or 9 _____

14 -13 or 5 _____

15 -7 or 7 _____

16 18 or -8 _____

Find each product.

17 $6 \cdot (-7) =$ _____

18 $-9 \cdot (-8) =$ _____

19 $-12 \cdot 5 =$ _____

20 $10 \cdot (-11) =$ _____

Copyright © Glencoe/McGraw-Hill, a division of The McGraw-Hill Companies, Inc.

Divide.

21 $-144 \div 12 =$ _____

22 $\dfrac{-66}{-11} =$ _____

23 $-27 \div 1 =$ _____

24 $360 \div (-10) =$ _____

Solve.

25 **HEALTH** Collin went to the doctor. The cost was $164. His insurance was expected to pay $115. Collin paid $49. Later, Collin found out that his insurance actually paid $131. Assuming Collin had no prior balance, what is the balance after his insurance payment was received?

_____

26 **FINANCES** Amelia had $146 in her checking account. She used her debit card to purchase dinner for $17 and a birthday present for her mother for $28. On Friday, she deposited $75. What is her checking account balance after her deposit?

_____

_____

Correct the mistakes.

27 **CHEMISTRY** The boiling point of neon is approximately $-246°C$. The melting point of neon is about $-259°C$. Trent says that the difference between these two temperatures is -505. What is wrong with Trent's answer?

_____

_____

28 **WEATHER** The weather forecaster on the radio said, "It is currently five degrees below zero. We expect that the sun will warm up the temperatures this afternoon. The temperature should rise about thirty degrees during the day." Salvador told his friend that the high temperature that day would be 35°F. What is wrong with Salvador's answer?

_____

Copyright © Glencoe/McGraw-Hill, a division of The McGraw-Hill Companies, Inc.

Choose the best answer and fill in the corresponding circle on the sheet at right.

1 Due to a construction project, a small electronics store loses $3,000 per day in business. If this pattern continues for the next 4 days, how much money will this store have compared to its normal sales?

    A −$12,000   **C** −$8,000

    B −$10,000   **D** −$4,000

2 Which symbol makes this math sentence true?

$$-500 - 27 \bigcirc -500 - (-27)$$

    A <   **C** >

    B =   **D** −

3 A football team lost 39 yards on 3 plays. If the team lost the same number of yards on each play, which integer shows the yards lost per play?

    A −13 yards   **C** −10 yards

    B −12 yards   **D** −8 yards

4 Alex owed his sister $15. He also owed his father $22. If he earns $65 next week, how much money will he have left after he pays what he owes?

    A $28   **C** $72

    B $58   **D** $102

5 Which letter on the number line represents −3?

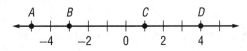

    A *A*   **C** *C*

    B *B*   **D** *D*

6 Toshiro went scuba diving. He dove to 75 feet below sea level to explore a cave in Manatee Springs. Then he climbed 39 feet to take pictures of jellyfish. Where is Toshiro in relation to sea level?

    A −74 feet   **C** 36 feet

    B −36 feet   **D** 74 feet

7 Find the product of −12 and −8.

    A 96   **C** −20

    B −4   **D** −96

8 What is the product of −108 and 4?

    A 432   **C** −104

    B −112   **D** −432

Copyright © Glencoe/McGraw-Hill, a division of The McGraw-Hill Companies, Inc.

9 Esmerelda has stock in the Capital Calendar Company. At the beginning of the day the price per share was $37.00. At the end of the day, the price per share had gone down by $3.00. What was the price per share at the end of the day?

A −$40

B $34

C $40

D $111

---

10 Find the quotient of −56 and −8.

A −64

B −48

C −7

D 7

---

11 What is the opposite of 8?

A −8

B 0

C $\frac{1}{8}$

D 1

---

12 Tamara and her family went hiking in the Appalachian Mountains. They descended the mountain at a rate of 6 feet per minute. What is their change in elevation after 15 minutes?

A −90

B −21

C 21

D 90

ANSWER SHEET

Directions: Fill in the circle of each correct answer.

1 Ⓐ Ⓑ Ⓒ Ⓓ

2 Ⓐ Ⓑ Ⓒ Ⓓ

3 Ⓐ Ⓑ Ⓒ Ⓓ

4 Ⓐ Ⓑ Ⓒ Ⓓ

5 Ⓐ Ⓑ Ⓒ Ⓓ

6 Ⓐ Ⓑ Ⓒ Ⓓ

7 Ⓐ Ⓑ Ⓒ Ⓓ

8 Ⓐ Ⓑ Ⓒ Ⓓ

9 Ⓐ Ⓑ Ⓒ Ⓓ

10 Ⓐ Ⓑ Ⓒ Ⓓ

11 Ⓐ Ⓑ Ⓒ Ⓓ

12 Ⓐ Ⓑ Ⓒ Ⓓ

Success Strategy

When checking your answers, do not change your mind on your answer choice unless you misread the question. Your first choice is often the right one.

Copyright © Glencoe/McGraw-Hill, a division of The McGraw-Hill Companies, Inc.

Chapter 2

Patterns and Graphs

Have you ever calculated driving distance?

The length of one lap on a certain race track is 2.5 miles. If a car is driving 125 miles per hour, how many laps will the driver complete in 15 minutes?

Copyright © Glencoe/McGraw-Hill, a division of the McGraw-Hill Companies, Inc.

STEP 1 Quiz

Math Online > Are you ready for Chapter 2? Take the Online Readiness Quiz at *glencoe.com* to find out.

STEP 2 Preview

Get ready for Chapter 2. Review these skills and compare them with what you will learn in this chapter.

What You Know	What You Will Learn
You know how to add and follow patterns.	*Lesson 2-1*

What You Know

You know how to add and follow patterns.

Example: Each shipping box holds 75 DVDs. How many DVDs can two shipping boxes hold?

$$75 + 75 = 150$$

What You Will Learn

Lesson 2-1

Arithmetic sequences follow rules.

The rule is "Each shipping box holds 75 DVDs."

1 shipping box	75 DVDs
2 shipping boxes	150 DVDs
3 shipping boxes	225 DVDs
4 shipping boxes	300 DVDs

Add 75 DVDs for each additional shipping box.

What You Know

You know how to add the same number several times.

Example: If 4 flowers cost $5, how much does it cost to buy 12 flowers?

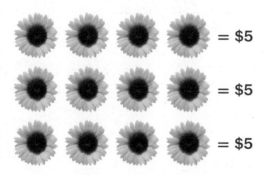

= $5

= $5

= $5

$$\$5 + \$5 + \$5 = \$15$$

What You Will Learn

Lesson 2-2

Patterns follow rules. You can use a table to describe a pattern.

Number of Flowers	4	8	12
Cost	$5	$10	$15

Hill, a division of The McGraw-Hill Companies, Inc.

Number Relationships

KEY Concept

Patterns follow a rule. You can use a **rule** to answer questions about the pattern and to predict **terms**.

Rules define relationships between numbers. For example, the length of the radius of a circle is half the length of its diameter.

The table below shows the relationship between the numbers.

Radius	$\frac{1}{2}$ in.	1 in.	$1\frac{1}{2}$ in.	2 in.	$2\frac{1}{2}$ in.
Diameter	1 in.	2 in.	3 in.	4 in.	5 in.

You can use number patterns to find the next three terms. A circle with a 6-inch diameter has a 3-inch radius. A circle with a 7-inch diameter has a $3\frac{1}{2}$-inch radius. A circle with an 8-inch diameter has a 4-inch radius.

VOCABULARY

pattern
  a sequence of numbers, figures, or symbols that follows a rule or design

rule
  tells how numbers are related to each other

term
  each number in a sequence

Sometimes patterns can follow rules with more than one operation.

Example 1

Esther had a balance of $25 in her checking account. Each day she withdrew $8 for lunch. What is the balance of her checking account after 5 days? Use two rules to find the answer.

1. One rule is add −$8 each day. Start with her account balance of $25.

> Remember that adding −8 is equal to subtracting 8.

Day 1 $25 + (-8) = 17$ Day 2 $17 + (-8) = 9$

Day 3 $9 + (-8) = 1$ Day 4 $1 + (-8) = -7$

Day 5 $-7 + (-8) = -15$

2. Another rule is multiply the number of days by $8, and then subtract.

$5 \cdot 8 = 40$ $25 - 40 = -15$

3. Esther has a balance of −$15 in her account.

Copyright © Glencoe/McGraw-Hill, a division of The McGraw-Hill Companies, Inc.

Copyright © Glencoe/McGraw-Hill, a division of The McGraw-Hill Companies, Inc.

YOUR TURN!

A workbook has 159 pages. Each lesson is 7 pages. If Dominick tears out 4 lessons, how many pages are left? Use two rules to find the answer.

1. One rule is _____.

   $159 -$ _____ $=$ _____   _____ $-$ _____ $=$ _____

   _____ $-$ _____ $=$ _____   _____ $-$ _____ $=$ _____

2. Another rule is _____.

   _____ $\cdot$ _____ $=$ _____   _____ $-$ _____ $=$ _____

3. There are _____ pages left in the book.

Example 2

Find a rule. Then write the next three terms.

0.75; 1.50; 2.25; 3.00; 3.75

1. One rule is add 0.75.

   $0.75 + 0.75 = 1.50$
   $1.50 + 0.75 = 2.25$
   $2.25 + 0.75 = 3.00$
   $3.00 + 0.75 = 3.75$

2. Continue the pattern.

   $3.75 + 0.75 = 4.50$
   $4.50 + 0.75 = 5.25$
   $5.25 + 0.75 = 6.00$

3. The next three terms are 4.50; 5.25; and 6.00.

YOUR TURN!

Find a rule. Then write the next three terms.

$$\frac{1}{12}, \frac{3}{12}, \frac{5}{12}, \frac{7}{12}, \frac{9}{12}$$

1. One rule is _____.

   $\frac{1}{12}$ _____ $=$ _____

   _____ $=$ _____

   _____ $=$ _____

   _____ $=$ _____

2. Continue the pattern.

   _____

   _____

   _____

3. The next three terms are _____,

   _____, and _____.

GO ON

Example 3

Find a rule. Then write the next three terms.

1, 6, 31, 156

1. One rule is multiply by 5, and then add 1.

   $1 \cdot 5 = 5 + 1 \quad = 6$
   $6 \cdot 5 = 30 + 1 \quad = 31$
   $31 \cdot 5 = 155 + 1 = 156$

2. Continue the pattern.

   $156 \cdot 5 = 780 + 1 \quad = 781$
   $781 \cdot 5 = 3{,}905 + 1 \; = 3{,}906$
   $3{,}906 \cdot 5 = 19{,}530 + 1 = 19{,}531$

The next three terms are 781; 3,906; 19,531.

YOUR TURN!

Find a rule. Then write the next three terms.

4; 34; 334; 3,334

1. One rule is _____

   _____

   _____

   _____

   _____

2. Continue the pattern.

   _____

   _____

   _____

The next three terms are _____;

_____; _____.

Who is Correct?

How many ounces are in 5 quarts?

Number of Quarts	1	2	3	4	5
Number of Ounces	32	64			

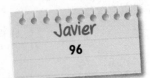
Javier
96

Rojen
160

Roxanne
150

Circle correct answer(s). Cross out incorrect answer(s).

▶ Guided Practice

Find a rule for each pattern.

1 207, 202, 197, 192 _____

2 124, 141, 158, 175 _____

3 117,649; 16,807; 2,401; 343 _____

4 39; 117; 351; 1,053 _____

Copyright © Glencoe/McGraw-Hill, a division of The McGraw-Hill Companies, Inc.

5 There are 48 apples in each crate. Find a rule. Then write the next three terms in the pattern.

Step 1 One rule is multiply by _____.

Step 2 Multiply the number of crates by _____ to continue the pattern.

2 · _____ = _____   3 · _____ = _____   4 · _____ = _____

Step 3 The next three terms are _____, _____, and _____.

In each sequence, find a rule. Then write the next three terms.

6 479, 456, 433, 410, _____, _____, _____

Rule: _____

410 − _____ = _____

_____ − _____ = _____

_____ − _____ = _____

The next three terms are _____, _____, and _____.

7 12,672; 6,336; 3,168; 1,584; _____; _____; _____

Rule: _____

1,584 ÷ _____ = _____

_____ ÷ _____ = _____

_____ ÷ _____ = _____

The next three terms are _____; _____; and _____.

8 57; 228; 912; 3,648; _____; _____; _____

Rule: _____

The next three terms are _____; _____; and _____.

9 153; 384; 846; 1,770; _____; _____; _____

Rule: _____

The next three terms are _____; _____; and _____.

GO ON

Copyright © Glencoe/McGraw-Hill, a division of The McGraw-Hill Companies, Inc.

Write the next three conversions in each pattern.

10

Number of Yards	1	2	3	4
Number of Inches	36			

11

Number of Miles	1	2	3	4
Number of Yards	1,760			

Step by Step Problem-Solving Practice

Solve.

Problem-Solving Strategies

☑ Make a table.
☐ Guess and check.
☐ Act it out.
☐ Solve a simpler problem.
☐ Work backward.

12 FASHION Marcella bought 7 sweaters. The first sweater cost $36.75. Each additional sweater cost $28.25. How much did she spend in all?

Understand Read the problem. Write what you know.

The first sweater cost _____.

Each additional sweater cost _____.

Plan Pick a strategy. One strategy is to make a table. Label the rows of the table "Sweater" and "Cost."

Solve One sweater cost _____. Each

additional sweater costs _____.

The rule is _____.

To find the total cost, complete the table.

Sweater	1	2	3	4	5	6	7
Cost							

Marcella spent _____ on sweaters.

Check Use a different rule to check your answer. Start with your answer and work backward.

Copyright © Glencoe/McGraw-Hill, a division of The McGraw-Hill Companies, Inc.

13 PETS Manuel is buying an aquarium. The store will charge $579.95 for the aquarium and an additional $30.75 for each fish. What is the price for the aquarium and 8 fish? Check off each step.

_____ Understand: I underlined key words.

_____ Plan: To solve the problem I will _____.

_____ Solve: The answer is _____.

_____ Check: I checked my answer by _____.

14 ZOOS The zookeeper ordered 300 kilograms of feeding fish. Each day, Sarah the sea lion eats 17.4 kilograms of fish. After 14 days, how many kilograms of fish will remain?

_____

_____

15 Reflect Explain a rule for the terms 32, 69, 143, 291.

_____

_____

▶ Skills, Concepts, and Problem Solving

Find a rule for each pattern.

16 248,832; 20,736; 1,728; 144 _____

17 845.6; 837.7; 829.8; 821.9 _____

18 $\frac{9}{12}$; $1\frac{2}{12}$; $1\frac{7}{12}$; 2 _____

19 2; 13; 46; 145 _____

In each sequence, find a rule. Then write the next three terms.

20 2, −4, 8, −16, 32

Rule: _____

Next terms: _____; _____; _____

21 781,250; 156,250; 31,250; 6,250

Rule: _____

Next terms: _____; _____; _____

22 641.72; 650.00; 658.28; 666.56

Rule: _____

Next terms: _____; _____; _____

23 3, 12, 30, 66

Rule: _____

Next terms: _____; _____; _____

GO ON

Copyright © Glencoe/McGraw-Hill, a division of The McGraw-Hill Companies, Inc.

Write the next three conversions in each pattern.

24

Fluid Ounces	1	2	3	4	5
Gallons	128				

25

Days	1	2	3	4	5
Minutes	1,440				

Solve.

26 **FINANCE**   Sergei opened his savings account with $467. Each week he withdrew $62. What is the balance in his savings account after 7 weeks?

_____

27 **THEATERS**   A new stadium-style movie theater just opened in town. The first row has 16 seats. Each row after has two additional seats. How many seats are 8 rows of the theater?

Rows	1	2	3	4	5	6	7	8
Total Number of Seats	16	34	54					

Vocabulary Check   **Write the vocabulary word that completes each sentence.**

28 The number 396 is a(n) _____ in the sequence 300, 396, 492, 588.

29 A _____ is a sequence of numbers, figures, or symbols that follows a rule or design.

30 **Writing in Math**   Compare and contrast the differences between these two sequences.

Sequence A: 1, −3, 9, −27, 81
Sequence B: 1, 3, 9, 27, 81

_____

_____

_____

_____

STOP

Copyright © Glencoe/McGraw-Hill, a division of The McGraw-Hill Companies, Inc.

Introduction to Functions

KEY Concept

A **function** is a relationship in which one quantity depends upon another quantity. The function assigns exactly one output value to each input value using a rule.

A **function table** uses the function, or rule, to show the relationship between the values. The function is written in an equation.

Imagine that a coral reef grows 2 millimeters each year. The value y shows the amount of change in reef size each year.

growth of the coral reef — **Function:** $y = 2x$ — number of years

Function Table		
Input	Function	Output
x	$2x$	y
0	2 • 0	0
1	2 • 1	2
2	2 • 2	4
3	2 • 3	6

VOCABULARY

equation
a mathematical sentence that contains an equal sign

function
a relationship in which one quantity depends upon another quantity (for every x-value there is exactly one y-value)

function table
a table of ordered pairs that is based on a rule

variable
a symbol, usually a letter, used to represent a number

The function table above shows the growth of the coral reef, y, based on the number of years, x.

Example 1

Write a function to represent the situation.

Jodi is 6 years older than Max.

1. Let x = Max's age and y = Jodi's age.

2. The function is $y = x + 6$.

YOUR TURN!

Write a function to represent the situation.

Every triangle has 3 sides.

1. Let ____ = the number of triangles and ____ = the number of sides.

2. The function is _____ .

Copyright © Glencoe/McGraw-Hill, a division of The McGraw-Hill Companies, Inc.

GO ON

Example 2

Make a function table for the equation y = −2x + 2.

1. Substitute each value of x in the function.

2. Solve the equation to find y.

x	−2	−1	0	1	2
y = −2x + 2	(−2) (−2) + 2	(−2) (−1) + 2	(−2) (0) + 2	(−2) (1) + 2	(−2) (2) + 2
y	6	4	2	0	−2

YOUR TURN!

Make a function table for the equation y = 2x + 1.

1. Substitute each value of x in the function.

2. Solve the equation to find y.

x	−2	−1	0	1	2
y = 2x + 1					
y					

Who is Correct?

Make a function table for the equation y = x − 5.

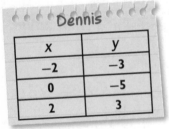

Dennis

x	y
−2	−3
0	−5
2	3

Quinton

x	y
−2	−7
0	5
2	3

Polly

x	y
−2	−7
0	−5
2	−3

Circle correct answer(s). Cross out incorrect answer(s).

Copyright © Glencoe/McGraw-Hill, a division of The McGraw-Hill Companies, Inc.

Guided Practice

Write a function to represent each situation.

1 Mike walks 2 miles fewer than Darrell every day. _____

2 Jarrod feeds 4 times as many fish as Patricia.   _____

Step by Step Practice

Write a function and make a function table.

3 MUSIC   A music club charges $3 for membership and $2 for every downloaded song. How much will Alonso pay for joining the music club and downloading 6 songs?

Step 1   Write a function to describe the cost of membership and x downloads. Let $x =$ the number of downloaded songs and $y =$ cost.

$y =$ _____

Step 2   Make a function table using the rule $y =$ _____.

Number of Downloads, x	1	2	3	4	5	6
Total Cost, y						

Step 3   Alonso will pay $_____ for joining the music club and downloading 6 songs.

Write a function and make a function table.

4 SCHOOL   Each book Jackie carries weighs 3 pounds. Her backpack weighs 1 pound. Write a function to show the weight of her backpack if she is carrying x books.

$y =$ _____

Number of Books, x	1	2	3	4
Weight, y				

If Jackie carries 3 books, her backpack weighs _____ pounds.

Step by Step Problem-Solving Practice

Write a function and make a function table.

Problem-Solving Strategies
- ☐ Draw a diagram.
- ☑ Look for a pattern.
- ☐ Guess and check.
- ☐ Act it out.
- ☐ Work backward.

5 **SCIENCE** Martin is learning about insects, which have exactly 6 legs. He found a picture of four insects. How many legs would Martin have counted?

Understand Read the problem. Write what you know.

Insects have _____ legs.

Plan Pick a strategy. One strategy is to look for a pattern. Write a function to describe the situation.

Solve Let x = the number of insects and y = the number of legs.

SCIENCE Martin is learning about insects.

Make a function table using the rule $y =$ _____.

Number of insects, x	1	2	3	4
Legs, y				

Martin counted _____ insect legs.

Check Substitute _____ for x in the function.

6 **CELEBRATIONS** A high-school reunion is celebrated every 5 years. Keanu was 18 years old when he graduated. How old will he be at his third high-school reunion? Check off each step.

_____ Understand: I underlined key words.

_____ Plan: To solve this problem, I will _____.

_____ Solve: The function is _____.

_____ Check: I checked my answer by _____.

Reunion, x	1	2	3	4
Keanu's age, y				

Keanu will be _____ years old.

Copyright © Glencoe/McGraw-Hill, a division of The McGraw-Hill Companies, Inc.

7 **Reflect** Does $y = 4x - 3$ match the data in the table? Explain.

x	−2	−1	0	1	2
y					

Skills, Concepts, and Problem Solving

Write a function to represent each situation.

8 Linda writes 3 fewer e-mails than Suja each day. _____

9 Bruce works 5 times as many hours as Ronnie. _____

Write a function and make a function table.

10 **TRAVEL** Kala drove from Detroit to New York City. She drove 45 miles every hour. If it took Kala 15 hours to drive from Detroit to New York City, how many miles did Kala drive?

$y = $ _____

Number of Hours, x	3	6	9	12	15
Number of Miles, y					

Kala drove _____ miles to get to New York City.

11 **FAMILY** Keenan is 3 years older than his sister Tracey. How old will Keenan be when Tracey is 4?

$y = $ _____

Tracey's Age, x		Keenan's Age, y
1	+ 3	
2	+ 3	
3	+ 3	
4	+ 3	

Keenan will be _____ when Tracey is 4.

GO ON

Write a function and make a function table.

12 **CHEMISTRY**   In a lab experiment, a scientist used 2.5 liters of solution for every milliliter of water. If the scientist had to mix 8 milliliters of water, how many liters of solution will the scientist need?

$y =$ _____

Milliliters of Water, x	2	4	6	8
Liters of Solution, y				

The scientist needs _____ liters of solution.

Vocabulary Check   **Write the vocabulary word that completes each sentence.**

13 A(n) _____ is a table of ordered pairs that is based on a rule.

14 A(n) _____ is a relationship in which one quantity depends upon another quantity.

15 **Writing in Math**   Explain how to make a function table.

_____

_____

▶ **Spiral Review**

16 **FITNESS**   Courtney practices soccer for 45 minutes each day except for Sundays. After 2 weeks, how much time will Courtney have practiced? (Lesson 2-1, p. 50)

_____

_____

STOP

Copyright © Glencoe/McGraw-Hill, a division of The McGraw-Hill Companies, Inc.

Progress Check 1 (Lessons 2-1 and 2-2)

In each sequence, find a rule. Then write the next three terms.

1   4, −12, 36, −108, 324

Rule: _____

Next terms: _____; _____; _____

2   250; 247.7; 245.4; 243.1

Rule: _____

Next terms: _____; _____; _____

Write a function to represent each situation.

3   Hannah earns $8 an hour more than Russ. _____

4   A quilting club charges $50 for membership and $100 for every quilt purchased.

_____

Write a function and make a function table.

5   **CRAFTS**   Aurelia knits approximately 5 new scarves each week. In 6 weeks, about how many scarves would Aurelia make?

$y = $ _____

Number of Weeks, x	1	2	3	4	5	6
Number of Scarves, y						

Aurelia would make about _____ scarves in 6 weeks.

6   Juan counted the number of cars that pass his house each hour. How many cars have passed after 4 hours?

$y = $ _____

Hours, x	1	2	3	4
Cars Passed, y				

After 4 hours, _____ cars have passed.

Copyright © Glencoe/McGraw-Hill, a division of The McGraw-Hill Companies, Inc.

Ordered Pairs

KEY Concept

Follow these steps to graph the **ordered pair** (4, 5) on a **coordinate grid**.

> The *x*-coordinate is the first part of an ordered pair. It indicates how far to the left or to the right of the *y*-axis the corresponding point is located.

> The *y*-coordinate is the second part of an ordered pair. It indicates how far above or below the *x*-axis the corresponding point is located.

1. Start at the origin, (0, 0).

2. Move 4 units to the right, along the *x*-axis.

3. Then move 5 units up, along a line parallel to the *y*-axis.

4. Plot a point.

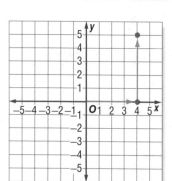

VOCABULARY

coordinate grid
a grid in which a horizontal number line and a vertical number line intersect at their zero points

ordered pair
a pair of numbers that are the coordinates of a point in a coordinate grid written in the order (horizontal coordinate, vertical coordinate)

origin
the point (0, 0) on a coordinate graph where the vertical axis crosses the horizontal axis

x-axis
the horizontal axis (↔) in a coordinate graph

y-axis
the vertical axis (↕) in a coordinate graph

Start at the origin, (0, 0). If the *x*-coordinate is positive, move to the *right*. If it is negative, move to the *left*. If the *y*-coordinate is positive, move *up*. If it is negative, move *down*.

Example 1

Name the ordered pair for point R.

1. Start at the origin, (0, 0).

2. Move to the right, along the *x*-axis, until you are above point R. You moved right 5 units, so the *x*-coordinate is 5.

3. Then move 3 units down, along a line parallel to the *y*-axis, until you reach point R. You moved down 3 units, so the *y*-coordinate is −3.

4. The ordered pair for point R is (5, −3).

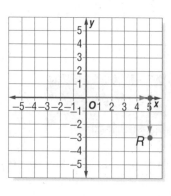

Copyright © Glencoe/McGraw-Hill, a division of The McGraw-Hill Companies, Inc.

Copyright © Glencoe/McGraw-Hill, a division of The McGraw-Hill Companies, Inc.

YOUR TURN!

Name the ordered pair for point _T_.

1. Start at the origin, (0, 0).

2. Move to the _____, along the _x_-axis, until you are under point _T_. You moved _____ _____ units, so the _x_-coordinate is _____.

3. Then move _____ units _____ along a line parallel to the _y_-axis, until you reach point _T_. You moved _____ _____ units, so the _y_-coordinate is _____.

4. The ordered pair for point _T_ is _____.

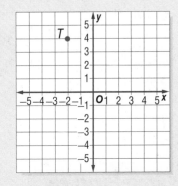

Example 2

Graph (−2, −4).

1. Start at the origin, (0, 0).

2. Move 2 units to the left, along the _x_-axis.

3. Then move 4 units down, along a line parallel to the _y_-axis. Plot a point.

YOUR TURN!

Graph (5, −3).

1. Start at the origin, (0, 0).

2. Move _____ units to the _____, along the _x_-axis.

3. Then move _____ units _____, along a line parallel to the _y_-axis. Plot a point.

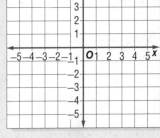

Who is Correct?

Name the ordered pair for point _A_.

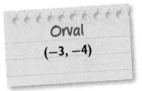

Orval
(−3, −4)

Lonzo
(3, −4)

Nicole
(3, 4)

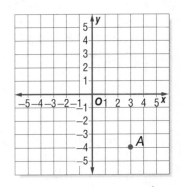

Circle correct answer(s). Cross out incorrect answer(s).

GO ON

Guided Practice

Name the ordered pair for each point.

1 D _____

2 E _____

3 F _____

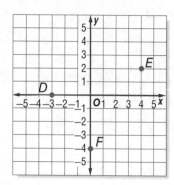

Step by Step Practice

4 Graph the ordered pairs $A(4, 3)$ and $B(-5, 3)$. Then connect the points. What do you notice?

Step 1 Graph point A. Start at the origin, $(0, 0)$. Move _____ units to the _____, along the x-axis. Then move _____ units _____, along a line parallel to the y-axis. Plot a point.

Step 2 Graph point B. Start at the origin, $(0, 0)$. Move _____ units to the _____, along the x-axis. Then move _____ units _____, along a line parallel to the y-axis. Plot a point.

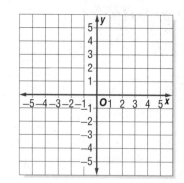

Step 3 Connect the points with a line. $(4, 3)$ and $(-5, 3)$ are on a line parallel to the x-axis because they have the same _____-coordinate.

Graph the ordered pairs.

5 Graph the ordered pairs $M(-5, 2)$ and $N(-5, -4)$.

Graph point M. Start at the origin, $(0, 0)$.

Move _____ units to the _____. Then move

_____ units _____. Plot a point.

Graph point N. Start at the origin, $(0, 0)$. Move

_____ units to the _____. Then move

_____ units _____. Plot a point.

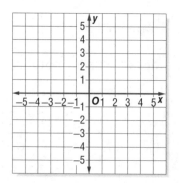

Copyright © Glencoe/McGraw-Hill, a division of The McGraw-Hill Companies, Inc.

6 Graph the ordered pairs $S(-1, -4)$ and $T(5, -4)$. Then connect the points. What do you notice?

$(-1, -4)$ and $(5, -4)$ have the same _____-coordinate.

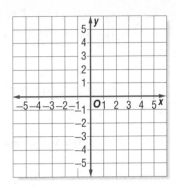

Step by Step Problem-Solving Practice

Problem-Solving Strategies
☐ Look for a pattern.
☐ Guess and check.
☐ Act it out.
☑ Use a graph.
☐ Work backward.

Solve.

7 **TRAVEL** Rina has a map of historical sites in her city. How many miles separate the museum from the town square? Each unit on the x- or y-axis represents one mile.

Understand Read the problem. Write what you know. You must find the distance between the

_____ and the _____.

Plan Pick a strategy. One strategy is to use a graph. Find the ordered pairs for the town square and the museum. Then find the distance from one to the other.

Solve The town square is located at _____.

The museum is located at _____.

The town square and the museum are on a line parallel to the x-axis. Count the number of units between the town square and the museum.

There are _____ units between the town square and the museum.

The museum is _____ miles from the town square.

Check Use the absolute value of the _____ -coordinates to verify the distance.
$|-1| + |5| = $ _____

> Remember:
> The absolute value is the distance between a number and 0 on a number line.

GO ON

Copyright © Glencoe/McGraw-Hill, a division of The McGraw-Hill Companies, Inc.

CITIES Reagan has a map of a city. Each unit on the x- or y-axis represents one block. Use the map to answer Exercises 8 and 9.

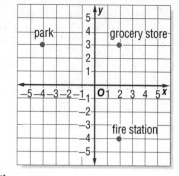

8 How many blocks separate the park from the grocery store? Check off each step.

_____ **Understand: I underlined key words.**

_____ **Plan: To solve this problem, I will** _____.

_____ **Solve: The answer is** _____.

_____ **Check: I checked my answer by** _____

_____.

9 How many blocks separate the grocery store from the fire station?

_____

10 Make a map of Satinka's school. Graph the ordered pairs.

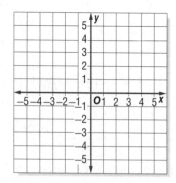

Cafeteria   (3, 4)
Classroom   (3, −2)
Library   (−4, 4)

What is the distance between the cafeteria and the library?

_____

11 **Reflect** How can you find the distance between two points on a coordinate graph with the same x- or y-coordinates?

_____

_____

_____

Copyright © Glencoe/McGraw-Hill, a division of The McGraw-Hill Companies, Inc.

▶ Skills, Concepts, and Problem Solving

Name the ordered pair for each point.

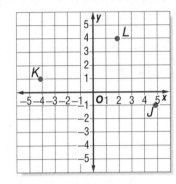

12 J _____

13 K _____

14 L _____

Graph the ordered pairs.

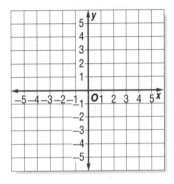

15 $Q(5, 4)$

16 $R(0, -3)$

17 $S(-4, 2)$

Solve.

TOWNS Morena has a map of a town. Each unit on the *x*- or *y*-axis represents one mile. Use the map to answer Exercises 18 and 19.

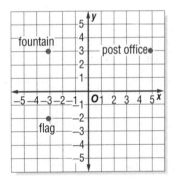

18 How many miles separate the fountain from the flag?

_____

19 How many miles separate the fountain from the post office?

_____

20 Make a map of Starrtown. Graph the ordered pairs.

Post Office   (2, 1)
School   (2, −3)
Grocery Store   (−1, 1)

What is the distance between the post office and the school?

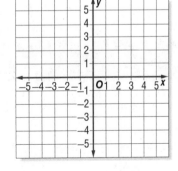

Copyright © Glencoe/McGraw-Hill, a division of The McGraw-Hill Companies, Inc.

GO ON

Write the vocabulary word that completes each sentence.

21 A(n) _____ is a pair of numbers that gives the location of a point on a map or coordinate grid.

22 The _____ is the point (0, 0) on a coordinate grid where the x-axis and y-axis intersect.

23 **Writing in Math** Explain how to graph the point (−3, −2) in two different ways.

_____

_____

▶ Spiral Review

Write a function to represent each situation. (Lesson 2-2, p. 57)

24 Dalila walks 3 more miles a day than Basilio.

_____

25 Aidia watches 2 times as many movies each week as Bob.

_____

26 Make a function table for the equation $y = 3x - 1$.

x	−2	−1	0	1	2
y = 3x − 1					
y					

Write a function and make a function table. (Lesson 2-2, p. 57)

27 **COMMUTING** Every day Victor travels 27.4 miles to and from work. In a five-day work week, how many miles does Victor travel?

$y = $ _____

Number of Days, x	1	2	3	4	5
27.4x					
Number of Miles, y					

Victor drives _____ miles.

STOP

Copyright © Glencoe/McGraw-Hill, a division of The McGraw-Hill Companies, Inc.

Coordinate Grids

KEY Concept

To graph an equation, substitute different x-values into the equation. Evaluate the equation to find the y-values.

The x and y values form an ordered pair. Graph the ordered pairs on a coordinate grid.

Equation

$$y = -x - 1$$

> The x values can be selected at random.

Table

x	$-x - 1$	y	Ordered Pair (x, y)
2	$-2 - 1$	-3	$(2, -3)$
1	$-1 - 1$	-2	$(1, -2)$
0	$0 - 1$	-1	$(0, -1)$
-1	$-(-1) - 1$	0	$(-1, 0)$
-2	$-(-2) - 1$	1	$(-2, 1)$

After you graph the ordered pairs, connect the points with a line.

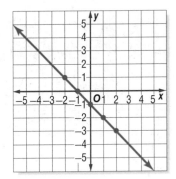

The ordered pair for each point on the line is a solution to the equation. This shows that the equation has an infinite number of solutions.

VOCABULARY

coordinate grid
a grid in which a horizontal number line and a vertical number line intersect at their zero points

ordered pair
a pair of numbers that are the coordinates of a point in a coordinate grid written in the order (horizontal coordinate, vertical coordinate)

Copyright © Glencoe/McGraw-Hill, a division of The McGraw-Hill Companies, Inc.

GO ON

Example 1

Graph the equation $y = x + 3$.

1. Make a table. Substitute $-2, -1, 0, 1,$ and 2 for x. Solve for y.

x	x + 3	y	Ordered Pair
−2	−2 + 3	1	(−2, 1)
−1	−1 + 3	2	(−1, 2)
0	0 + 3	3	(0, 3)
1	1 + 3	4	(1, 4)
2	2 + 3	5	(2, 5)

2. Graph the ordered pairs. Connect the points with a line.

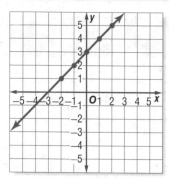

YOUR TURN!

Graph the equation $y = -2x - 3$.

1. Make a table. Substitute $-3, -2, -1, 0,$ and 1 for x. Solve for y.

x	−2x − 3	y	Ordered Pair
−3	−2(−3) − 3	3	(−3, 3)
−2	−2(−2) − 3	1	(−2, 1)
−1	−2(−1) − 3		
0			
1			

2. Graph the ordered pairs. Connect the points with a line.

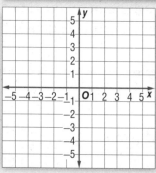

Copyright © Glencoe/McGraw-Hill, a division of The McGraw-Hill Companies, Inc.

Example 2

There are 3 feet in 1 yard. Show the relationship between the number of feet and the number of yards on a coordinate grid. How many feet are in 4 yards?

Let x = number of yards and y = number of feet.

feet	equals	3	times	yards
y	=	3	·	x

1. The equation is $y = 3x$.

2. Make a table. Substitute 0, 1, 2, 3, and 4 for x. Solve for y.

x	$3x$	y	Ordered Pair
0	3(0)	0	(0, 0)
1	3(1)	3	(1, 3)
2	3(2)	6	(2, 6)
3	3(3)	9	(3, 9)
4	3(4)	12	(4, 12)

3. Graph the ordered pairs. Connect the points with a line.

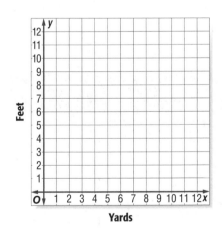

The ordered pair (4, 12) means that 4 yards have 12 feet.

YOUR TURN!

Maria earns 2 points for each layup she makes in the basketball game. Show the relationship between the number of layups she makes and the number of points she earns on a coordinate grid. How many points will Maria earn for making 5 layups?

Let x = number of layups made and y = number of points earned.

points earned	equals	2	times	layups
y	=	2	·	x

1. The equation is $y = 2x$.

2. Make a table. Substitute 1, 2, 3, 4, and 5 for x. Solve for y.

x	$2x$	y	Ordered Pair
1	2(1)		
2	2(2)		
3	2(3)		
4	2(4)		
5	2(5)		

3. Graph the ordered pairs. Connect the points with a line.

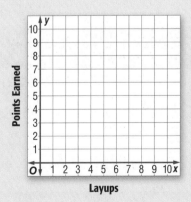

The ordered pair _____ means that Maria earned _____ points for making 5 layups.

GO ON

Copyright © Glencoe/McGraw-Hill, a division of The McGraw-Hill Companies, Inc.

Who is Correct?

Make a table for three ordered pairs for the equation $y = 3x - 7$.

Morton

x	3x − 7	y	Ordered Pair
−4	3(−4) − 7	−18	(−4, −18)
−1	3(−1) − 7	−10	(−1, −10)
0	3(0) − 7	−7	(0, −7)

Kiele

x	3x − 7	y	Ordered Pair
1	3(1) − 7	−4	(1, −4)
3	3(3) − 7	2	(3, 2)
4	3(4) − 7	5	(4, 5)

Ernesto

x	3x − 7	y	Ordered Pair
−2	3(−2) − 7	−13	(−2, −13)
0	3(0) − 7	−7	(0, −7)
2	3(2) − 7	−1	(2, −1)

Circle correct answer(s). Cross out incorrect answer(s).

 Guided Practice

Make a table for each equation.

1 $y = 9x - 4$

x	9x − 4	y	Ordered Pair
−2	9(−2) − 4	−22	(−2, −22)
−1	9(−1) − 4		
0			
1			
2			

2 $y = \frac{x}{2} + 6$

x	$\frac{x}{2} + 6$	y	Ordered Pair
−4	$\frac{-4}{2} + 6$		
−2	$\frac{-2}{2} + 6$		
0			
2			
4			

Copyright © Glencoe/McGraw-Hill, a division of The McGraw-Hill Companies, Inc.

3 Make a table for the equation.
$y = -x + 3$

x	$-x + 3$	y	Ordered Pair
-1	$-(-1) + 3$		
1	$-(1) + 3$		
2	$-(2) + 3$		
0	$-(0) + 3$		

4 Graph the equation from Exercise 3.

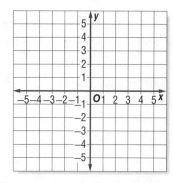

Step by Step Practice

5 Kyle pays $4d - 2$ dollars a week to join an exercise class at his health club, where d is the number of days he attends. Show the relationship between the number of days Kyle exercises at the health club, and the amount of money he will pay each week, on the coordinate grid. How much money, m, will Kyle pay if he attends the class 3 days this week?

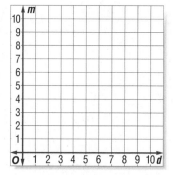

Step 1 Make a table. Substitute 1, 2, and 3 for d. Solve for m.

Step 2 Graph the ordered pairs. Connect the points with a line.

d	$4d - 2$	m	Ordered Pair
1	$4(1) - 2$		
2	$4(2) - 2$		
3	$4(3) - 2$		

The ordered pair _____ means that Kyle will pay $_____ to attend the exercise class 3 days this week.

Copyright © Glencoe/McGraw-Hill, a division of The McGraw-Hill Companies, Inc.

GO ON

Step by Step Problem-Solving Practice

Solve.

6 VIDEO GAMES It costs $3 to rent new video games and $1 to rent older games. Sabrina has $10 to rent video games. Substitute three values in the equation $3x + y = 10$ to show how many new, x, and older games, y, Sabrina can rent for $10.

Problem-Solving Strategies
☑ Make a graph.
☐ Look for a pattern.
☐ Guess and check.
☐ Act it out.
☐ Solve a simpler problem.

Understand Read the problem. Write what you know. Let _____ represent the number of new video games and _____ represent the older games. Sabrina can spend _____. The values that you substitute for x and y must fit the equation _____.

Plan Pick a strategy. One strategy is to make a graph.

Solve Rewrite the equation by solving for y.

$$3x + y = 10$$
$$3x + y - 3x = 10 - 3x$$
$$y = 10 - 3x$$

Then make a table. Substitute 1, 2, and 3 for x. Solve for y.

x	$10 - 3x$	y	Ordered Pair
1	$10 - 3(1)$		
2	$10 - 3(2)$		
3	$10 - 3(3)$		

Older Games / New Games

Graph the ordered pairs. Explain the solutions.

Sabrina can rent 1 new and _____ older games for $10.
Sabrina can rent 2 new and _____ older games for $10.
Sabrina can rent _____ new and _____ older game for $10.

Check Use multiplication to check your answers.

Copyright © Glencoe/McGraw-Hill, a division of The McGraw-Hill Companies, Inc.

7 **NATURE** Julio wants to have a picnic with his friends at a state park. The cost of admission to the park is $2 per person, plus $1 for parking. Write an equation to represent the situation. Then substitute four values to find out how much Julio will have to pay. Make a table and then graph the ordered pairs. Check off each step.

_____ Understand: I underlined key words.

_____ Plan: To solve the problem I will _____ .

_____ Solve: The answer is found below.

_____ Check: I checked my answer by _____ .

x	2x + 1	y	Ordered Pair
1			

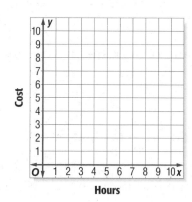

8 **Reflect** Explain how to graph the equation $y = 5x + 8$.

_____

_____

Skills, Concepts, and Problem Solving

9 Make a table for the equation.
$y = -3x + 4$

x	−3x + 4	y	Ordered Pair
−2	−3(−2) + 4	10	(−2, 10)
−1			
0			
1			
2			

10 Graph the equation from Exercise 9.

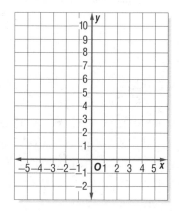

GO ON

Solve.

11 **JOGGING** Ronald walks 3 miles in 1 hour. Write an equation to represent the situation. Then substitute four values to find how far Ronald can walk in each situation. Make a table and then graph the ordered pairs to find the solutions.

h	3h	m	Ordered Pair
0	3(0)	0	(0, 0)
1			
2			
3			

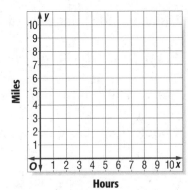

How many miles can Ronald walk in 3 hours?

_____

Vocabulary Check **Write the vocabulary word that completes each sentence.**

12 A(n) _____ is a grid in which a horizontal number line and a vertical number line intersect at their zero points.

13 A(n) _____ is a pair of numbers that are the coordinates of a point in a coordinate grid.

14 **Writing in Math** Explain how to make a table for the equation $y = -x + 7$.

_____

_____

_____

 Spiral Review

15 **SHAPES** Dylan wants to draw a square. The first three points are plotted on the coordinate grid. Where should the fourth point be plotted to make a square? Plot the point. (Lesson 2-3, p. 64) _____

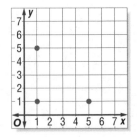

Copyright © Glencoe/McGraw-Hill, a division of The McGraw-Hill Companies, Inc.

Linear and Nonlinear Functions

KEY Concept

A **nonlinear function** is a set of ordered pairs that are related to each other by a non-constant rate. A **function table** can be used to create ordered pairs.

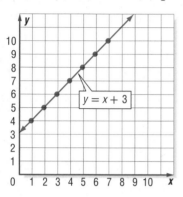

Linear Function

The **function** $y = x + 3$ is a straight line when graphed. The distance between each point along the x-axis is constant (the same). The distance between each point along the y-axis is constant.

Linear functions show a constant rate of change when graphed. So, $y = x + 3$ is a linear function.

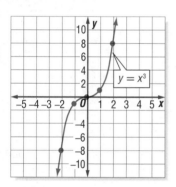

Nonlinear Function

The function $y = x^3$ is not a straight line when graphed. The distance between each point along the x-axis is constant. The distance between each corresponding point along the y-axis is not constant.

Nonlinear functions show a rate of change that is not constant when graphed. So, $y = x^3$ is a nonlinear function.

VOCABULARY

function
   a relationship in which one quantity depends upon another quantity

function table
   a table of ordered pairs that is based on a rule

linear function
   a function whose graph is a straight line

nonlinear function
   a function whose graph is not a straight line

The graph of a nonlinear function is not a straight line.

Copyright © Glencoe/McGraw-Hill, a division of The McGraw-Hill Companies, Inc.

Example 1

Match $y = -3x^2 - 1$ with its function table and its graph.

A.

x	−2	−1	0	1	2
y	13	4	0	4	13

B.

x	−2	−1	0	1	2
y	−13	−4	−1	−4	−13

I.
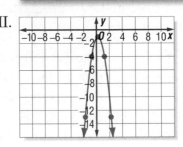

II.

1. Use a table to check the output values.

x	−2	−1	0	1	2
$-3x^2 - 1$	$-3(-2)^2 - 1$	$-3(-1)^2 - 1$	$-3(0)^2 - 1$	$-3(1)^2 - 1$	$-3(2)^2 - 1$
y	−13	−4	−1	−4	−13

2. Table B is the function table for $y = -3x^2 - 1$. Graph II is the graph of $y = -3x^2 - 1$.

YOUR TURN!

Match $y = 4x + 3$ with its function table and its graph.

A.

x	−2	−1	0	1	2
y	−5	−1	3	7	11

B.

x	−2	−1	0	1	2
y	−11	−7	3	7	11

I.

II.

1. Use a table to check the output values.

x	−2	−1	0	1	2
$4x + 3$					
y					

2. Table _____ is the function table for $y = 4x + 3$.
   Graph _____ is the graph of $y = 4x + 3$.

Copyright © Glencoe/McGraw-Hill, a division of The McGraw-Hill Companies, Inc.

Example 2

Make a function table and a graph for the function $y = x^2$. Is the function linear or nonlinear?

1. Make a function table using the rule $y = x^2$.

x	1	2	3	4	5
x^2	1^2	2^2	3^2	4^2	5^2
y	1	4	9	16	25

2. Graph the ordered pairs.

3. Evaluate the rate of change.

   If the rate of change is constant, draw a straight line.

   If the rate of change is not constant, draw a smooth curve to connect the points.

4. The function is nonlinear. The graph should show a smooth curve.

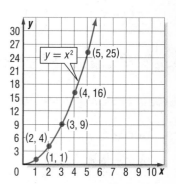

YOUR TURN!

Make a function table and a graph for the function $y = 2x^2$. Is the function linear or nonlinear?

1. Make a function table using the rule $y =$ _____.

x	1	2	3	4	5	6
$2x^2$						
y						

2. Graph the ordered pairs.

3. The rate of change is _____.

4. The function is _____.

   The graph should show a _____.

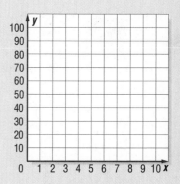

Copyright © Glencoe/McGraw-Hill, a division of The McGraw-Hill Companies, Inc.

GO ON

Who is Correct?

Make a function table for $y = -2x^3$.

Mindy

x	0	1	2
y	0	−2	16

RJ

x	0	1	2
y	0	−2	−12

Santos

x	0	1	2
y	0	−2	−16

Circle correct answer(s). Cross out incorrect answer(s).

 Guided Practice

Complete each function table.

1 $y = x^3 - 5$

x	−2	−1	0	1	2
y					

2 $y = 5x + 10$

x	−2	−1	0	1	2
y					

Match each function with its function table and its graph.

3 $y = 5x - 2$
Function table _____
Graph _____

4 $y = 2x^2 + 1$
Function table _____
Graph _____

A.

x	−2	−1	0	1	2
y	9	3	1	3	9

B.

x	−2	−1	0	1	2
y	−12	−7	−2	3	8

I.
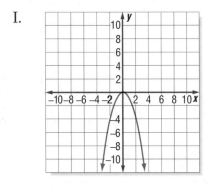

II.
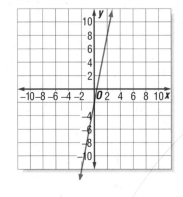

III.
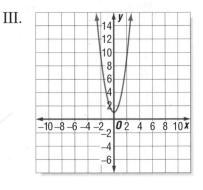

Copyright © Glencoe/McGraw-Hill, a division of The McGraw-Hill Companies, Inc.

5 Make a function table and a graph for the function $y = 100x^2$.
Is the function linear or nonlinear?

Step 1 Make a function table using the rule $y =$ _____.

x	1	2	3	4
y				

Step 2 Graph the ordered pairs.

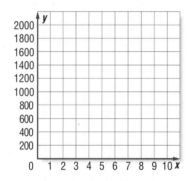

Step 3 The rate of change is _____.

Step 4 Connect the points. The function is _____.

The graph should show a _____.

6 Make a function table and a graph for $y = 3x$. Is the function
linear or nonlinear?

x	2	4	6	8	10
y					

The function is _____.

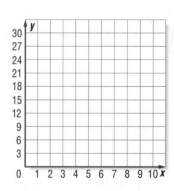

GO ON

Copyright © Glencoe/McGraw-Hill, a division of The McGraw-Hill Companies, Inc.

Problem-Solving Strategies
- ☑ Make a table.
- ☐ Look for a pattern.
- ☐ Guess and check.
- ☐ Act it out.
- ☐ Solve a simpler problem.

7 **INVESTING** Jada invested in a new company. The amount of money she earned is one dollar less than the cube of the number of years she has invested. How much money did Jada earn in the third year?

Understand Read the problem. Write what you know.

Write a function. Let x = the number of years and y = the number of dollars Jada earned.

Plan Pick a strategy. One strategy is to make a table.

Solve Make a function table using the rule $y = $ _____ .

Number of Years, x	1	2	3
Number of Dollars, y			

Graph the ordered pairs. Evaluate the rate of change. Connect the points.

In the third year, Jada earned _____ .

Check Substitute _____ for x in the function.

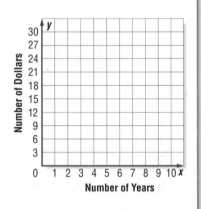

Copyright © Glencoe/McGraw-Hill, a division of The McGraw-Hill Companies, Inc.

8 **SAVING** Fay earns $3 per hour. How much will Fay earn after 5 hours of work? Check off each step.

_____ Understand: I underlined key words.

_____ Plan: To solve the problem I will _____ .

_____ Solve: I will use the rule _____ .

_____ Check: I will check my answer by _____ .

Number of Hours, x	1	2	3	4	5
Amount Earned, y					

Fay earns $_____ after 5 hours of work.

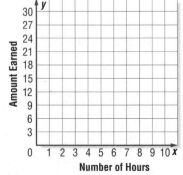

9 **Reflect** Does the function $y = 3x^3$ match the data in the function table? Explain.

x	−2	−1	0	1	2
y	−24	−3	0	3	9

_____

_____

 Skills, Concepts, and Problem Solving

Complete each function table.

10 $y = x + 7$

x	−2	−1	0	1	2
y					

11 $y = 4x^2$

x	−2	−1	0	1	2
y					

Match each function with its function table and its graph.

12 $y = -x$
Function table _____
Graph _____

13 $y = 3x^2 - 2$
Function table _____
Graph _____

A.

x	−2	−1	0	1	2
y	2	1	0	−1	−2

B.

x	−2	−1	0	1	2
y	10	1	−2	1	10

I.
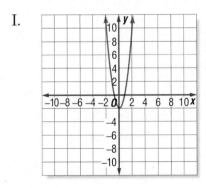

II.
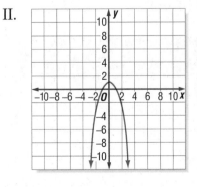

III.
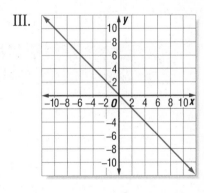

Copyright © Glencoe/McGraw-Hill, a division of The McGraw-Hill Companies, Inc.

GO ON

Make a function table and a graph for each function. Is the function linear or nonlinear?

14 $y = x(x + 8)$

x	1	2	3	4	5
y					

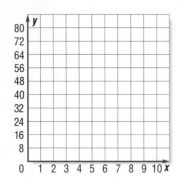

The function is _____.

15 $y = x^3 - 5$

x	2	3	4	5
y				

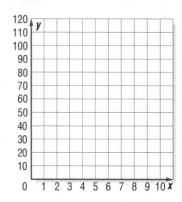

The function is _____.

Vocabulary Check   **Write the vocabulary word that completes each sentence.**

16 A(n) _____ is a relationship in which one quantity depends upon another quantity.

17 **Writing in Math**   Explain how to graph a nonlinear function.

_____

_____

▶ Spiral Review

18 Make a table for the equation $y = 4x + 2$. (Lesson 2-4, p. 71)

x	4x + 2	y	Ordered Pair
−2			
−1			
0			
1			
2			

STOP

Copyright © Glencoe/McGraw-Hill, a division of The McGraw-Hill Companies, Inc.

Name the ordered pair for each point.

1 A _____

2 B _____

3 C _____

4 D _____

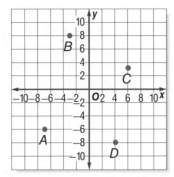

Write a function, make a function table, and make a graph. Is the function linear or nonlinear?

5 **CRAFTS** Carlita makes bracelets at Jade's Jewelry. She makes approximately 5 new bracelets to sell each day.

$y =$ _____

Number of Days, x	1	2	3	4	5	6
Number of Bracelets, y						

In 6 days, about how many bracelets would Carlita make to sell at the jewelry store?

Carlita would make about _____ bracelets in 6 days. The function is _____.

6 Jae measured how many cups of water drip from a broken sink faucet each hour.

$y =$ _____

Hours Passed, x	1	2	3	4
Cups of Water, y				

How many cups of water have dripped after 4 hours?

After 4 hours, _____ cups of water have dripped. The function is _____.

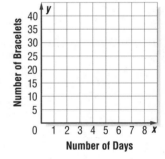

Copyright © Glencoe/McGraw-Hill, a division of The McGraw-Hill Companies, Inc.

Study Guide

Vocabulary and Concept Check

equation, *p. 57*

function, *p. 57*

function table, *p. 57*

linear function, *p. 79*

nonlinear function, *p. 79*

ordered pair, *p. 64*

origin, *p. 64*

pattern, *p. 50*

rule, *p. 50*

term, *p. 50*

variable, *p. 57*

x-axis, *p. 64*

y-axis, *p. 64*

Write the vocabulary word that completes each sentence.

1 A(n) _____ is a relationship in which one quantity depends upon another quantity.

2 A(n) _____ is a table of ordered pairs that is based on a rule.

3 A(n) _____ is a function whose graph is a line.

4 Each number in a sequence is called a(n) _____.

5 A mathematical sentence that contains an equal sign is called a(n) _____.

Label each diagram below. Write the correct vocabulary term in each blank.

6 _____

7 _____

8 _____

9 _____

10 _____

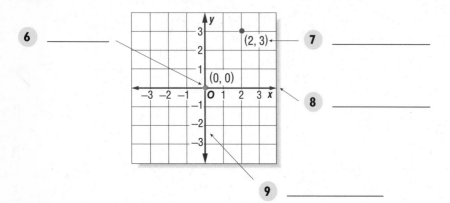

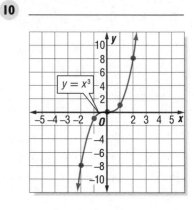

Lesson Review

2-1 Number Relationships (pp. 50–56)

Find a rule for each pattern.

11 185, 179, 173, 167 _____

12 −4, 8, −16, 32 _____

13 SHAPES   An octagon has 8 sides. How many sides do 7 octagons have?

Example 1

How many toes are on 6 owl feet?

1. Each foot has 4 toes. The rule is multiply the number of feet by 4.

2. There are 24 toes on 6 feet.

Copyright © Glencoe/McGraw-Hill, a division of The McGraw-Hill Companies, Inc.

2-2 Introduction to Functions (pp. 57-63)

Write a function and make a function table.

14 Mary is 3 years younger than Rae.

Let x = Rae's age and y = Mary's age.

$y =$ _____

Rae's Age, x	14	15	16	17	18	19
Mary's Age, y						

How old will Mary be when Rae is 19?

_____

Copyright © Glencoe/McGraw-Hill, a division of The McGraw-Hill Companies, Inc.

Example 2

Write a function and make a function table.

1. Every quadrilateral has 4 sides. What is the total number of sides in x quadrilaterals?

2. Use the function $y = 4x$ where x = the number of quadrilaterals and y = the number of sides.

3. Make a function table using the rule $y = 4x$.

Number of Quadrilaterals, x	1	2	3	4	5
Number of Sides, y	4	8	12	16	20

4. How many sides do 5 quadrilaterals have?
   Five quadrilaterals have a total of 20 sides.

2-3 Ordered Pairs (pp. 64-70)

Graph each ordered pair.

15 $A(-4, -4)$

16 $B(2, 4)$

17 $C(-3, 5)$

18 $D(5, 0)$

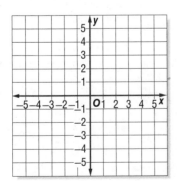

Example 3

Graph the ordered pair (5, −3).

1. Start at the origin, $(0, 0)$.
   Move 5 units to the right, along the x-axis.
   Then move 3 units down, along a line parallel to the y-axis.
2. Plot a point.

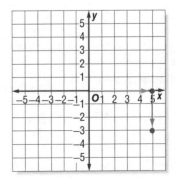

2-4 Coordinate Grids (pp. 71–78)

19 Graph the equation $y = -3x - 1$.

x	−3x − 1	y	Ordered Pair
−2	−3(−2) − 1	5	(−2, 5)
−1	−3(−1) − 1		
0	−3(0) − 1		
1	−3(1) − 1		

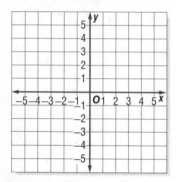

20 Tony has $3 in his savings account. He wants to save $2 more each week. Show the relationship between the number of dollars he saves each week and the total amount of money in his savings account, without interest.

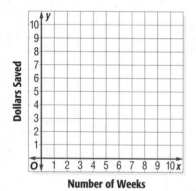

How much money will Tony have in his savings account after 3 more weeks?

Tony will have $_____.

Example 4

There are 2 pints in a quart. Write an equation to represent the situation. Then substitute five values to show the relationship between pints and quarts on a coordinate grid. How many pints are in 4 quarts?

1. Let x = number of pints and
   y = number of quarts.

quarts	equals	0.5	times	pints
y	$=$	0.5	$\cdot$	x

2. Write the equation: $y = 0.5x$.

3. Make a table. Substitute 0, 2, 6, 8, and 10 for x. Solve for y.

x	0.5x	y	Ordered Pair
0	0.5(0)	0	(0, 0)
2	0.5(2)	1	(2, 1)
6	0.5(6)	3	(6, 3)
8	0.5(8)	4	(8, 4)
10	0.5(10)	5	(10, 5)

4. Graph the ordered pairs. Connect the points with a line.

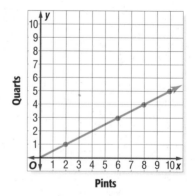

5. The ordered pair (10, 5) means that there are 10 pints in 5 quarts.

Copyright © Glencoe/McGraw-Hill, a division of The McGraw-Hill Companies, Inc.

2-5 Linear and Nonlinear Functions (pp. 79–86)

Write a function, make a function table, and make a graph. Is the function linear or nonlinear?

21 **VIDEOS** It will cost Peyton $15 to join the video game club. Then the club will charge him $4 for each video game rental.

$y = $ _____

Number of Video Games, x	1	2	3	4
Amount (in dollars), y				

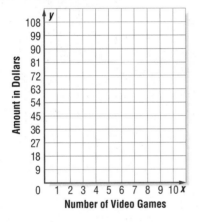

Number of Video Games

How much does Peyton pay if he joins the club and then rents 3 games?

Peyton pays $_____ for a membership plus $4 per video game rental.

The function is _____.

Example 5

Write a function, make a function table, and make a graph. Is the function linear or nonlinear?

GRADES For every A Jase earns on his report card, he receives three times the square of the number of A's in quarters from his grandfather. How many quarters does Jase receive if he earns 6 A's on his report card?

1. Use the function $y = 3x^2$ where $x =$ the number of A's on Jase's report card and $y =$ the amount received (in quarters).

2. Make a function table using the rule $y = 3x^2$.

Number of A's, x	1	2	3	4	5	6
Amount (in quarters), y	3	12	27	48	75	108

3. Graph the ordered pairs. Draw a smooth curve to connect the ordered pairs.

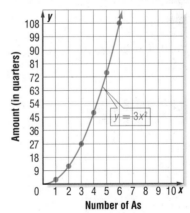

Number of As

4. Jase will receive 108 quarters, or $27, from his grandfather if he earns 6 A's on his report card. The function is nonlinear.

Copyright © Glencoe/McGraw-Hill, a division of The McGraw-Hill Companies, Inc.

Chapter Test

In each sequence, find a rule. Then write the next three terms.

1. 4, −12, 36, −108, 324

    Rule: _____

    Next terms:

    _____

2. 641.72; 627.50; 613.28; 599.06

    Rule: _____

    Next terms:

    _____

Write a function to represent each situation.

3. Tanner is 4 years older than Justice. _____

4. Zita spent $10 more than twice the amount Virginia spent.

    _____

Name the ordered pair for each point.

5. A _____

6. B _____

7. C _____

8. D _____

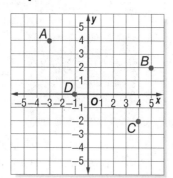

9. Make a table for the equation.
   $y = 3x - 2$

x	y
−1	
0	
1	
2	

10. Graph the equation from Exercise 9.

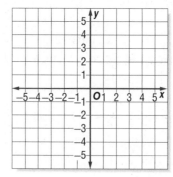

Copyright © Glencoe/McGraw-Hill, a division of The McGraw-Hill Companies, Inc.

Write a function, make a function table, and graph the ordered pairs.

11 **COOKING**   Diana is baking some cookies for her annual cookie exchange. She is going to make 4 batches of her cookies, and each batch requires 3 large eggs. How many eggs does she need to make x batches of cookies?

$y =$ _____

Number of Cookie Batches, x	1	2	3	4
Number of Eggs, y				

How many eggs would Diana use to make all 4 batches of her cookies? _____

Make a function table and graph for the function. Is the function linear or nonlinear?

12 $y = x^3 + 6$

x	1	2	3	4
y				

The function is _____.

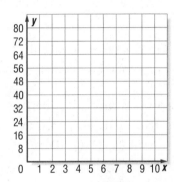

Solve.

13 **TRAVEL**   During each day of her one-week vacation on Mackinac Island, Monica biked a distance of 15 miles. Over the entire vacation, how many miles did Monica bike?

_____

Correct the mistakes.

14 On Nikki's math quiz, the problem stated: "For every mile you bicycle, you burn 35 Calories. Write the function."

What is the mistake Nikki made?

_____

_____

_____

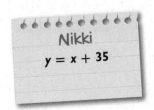

Nikki

$y = x + 35$

STOP

Copyright © Glencoe/McGraw-Hill, a division of The McGraw-Hill Companies, Inc.

Choose the best answer and fill in the corresponding circle on the sheet at right.

1 Which point on the grid corresponds to the ordered pair (4, 6)?

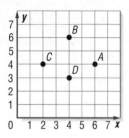

   A *A*

   B *B*

   C *C*

   D *D*

2 Which shape is made by plotting and connecting the following points on the coordinate grid: (3, 1), (7, 4), (7, 6), (3, 9)?

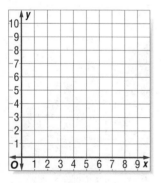

   A parallelogram

   B rectangle

   C square

   D trapezoid

3 Molly is selling T-shirts for a school fund-raiser. Use her chart to make a line graph that shows the data. If her goal is to raise at least $60, how many T-shirts does Molly need to sell?

T-Shirts Sold, x	Money Raised, y
1	$9
2	$18
3	$27

   A 4 T-shirts

   B 5 T-shirts

   C 6 T-shirts

   D 7 T-shirts

4 What is a rule for this pattern?

     30, 60, 120, 240, 480, 960

   A add 30

   B subtract 30

   C multiply by 2

   D add 60

Copyright © Glencoe/McGraw-Hill, a division of The McGraw-Hill Companies, Inc.

5 What is the next number in the sequence?

$$12, 17, 22, 27, 32, \underline{\hspace{1.5cm}}$$

A 37   C 42

B 40   D 47

6 Find the missing number.

x	5	8	9	12
y	10	16	18	?

A 27   C 25

B 26   D 24

7 Which ordered pair does *not* fall on the line for the equation $y = x + 3$?

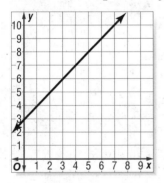

A (3, 7)   C (5, 8)

B (2, 5)   D (6, 9)

8 What is the next number in the sequence?

$$9, 18, 27, 36, 45, \underline{\hspace{1.5cm}}$$

A 50

B 54

C 58

D 60

ANSWER SHEET

Directions: Fill in the circle of each correct answer.

1 Ⓐ Ⓑ Ⓒ Ⓓ

2 Ⓐ Ⓑ Ⓒ Ⓓ

3 Ⓐ Ⓑ Ⓒ Ⓓ

4 Ⓐ Ⓑ Ⓒ Ⓓ

5 Ⓐ Ⓑ Ⓒ Ⓓ

6 Ⓐ Ⓑ Ⓒ Ⓓ

7 Ⓐ Ⓑ Ⓒ Ⓓ

8 Ⓐ Ⓑ Ⓒ Ⓓ

Success Strategy

When checking your answers, do not change your mind on your answer choice unless you misread the question. Your first choice is often the right one.

STOP

Copyright © Glencoe/McGraw-Hill, a division of The McGraw-Hill Companies, Inc.

Expressions and Equations

You solve equations all the time.

One night a weather forecaster says an additional 3 inches of snow fell, bringing the total snowfall for the year to 9 inches. You can solve the linear equation $x + 3 = 9$ to find the snowfall before the additional 3 inches fell.

Copyright © Glencoe/McGraw-Hill, Inc.

STEP 1 Quiz

Math Online ⟩ Are you ready for Chapter 3? Take the Online Readiness Quiz at *glencoe.com* to find out.

STEP 2 Preview

Get ready for Chapter 3. Review these skills and compare them with what you will learn in this chapter.

What You Know	What You Will Learn
You know how to add, subtract, multiply, and divide.	*Lesson 4-1*

What You Know

You know how to add, subtract, multiply, and divide.

Examples:

$3 + 4 = 7$ $45 \div 5 = 9$

TRY IT!

① $56 \div 7 =$ _____

② $57 - 18 =$ _____

③ $12 \cdot 4 =$ _____

④ $35 - 19 =$ _____

What You Will Learn

Lesson 4-1

When more than one operation is used in an equation, you must follow the **order of operations**.

1. Simplify the expressions inside grouping symbols.
2. Find the value of all powers.
3. Multiply and divide in order from left to right.
4. Add and subtract in order from left to right.

$$27 + 4^2 - (10 - 8) \cdot 3$$
$$= 27 + 4^2 - (10 - 8) \cdot 3$$
$$= 27 + 16 - 2 \cdot 3$$
$$= 27 + 16 - 6$$
$$= 43 - 6$$
$$= 37$$

You know how to use fact triangles to show relationships between numbers.

$8 + 6 = 14$
$6 + 8 = 14$
$14 - 8 = 6$
$14 - 6 = 8$

TRY IT!

⑤ _____

⑥ _____

⑦ _____

⑧ _____

Lesson 4-4

The relationship between **variables** can also be shown using fact triangles.

$r \cdot t = d$
$t \cdot r = d$
$d \div r = t$
$d \div t = r$

These equations show the relationship between distance, rate, and time. This is called the **distance formula**.

Copyright © Glencoe/McGraw-Hill, a division of The McGraw-Hill Companies, Inc.

Order of Operations

KEY Concept

You must follow the **order of operations** to evaluate mathematical expressions correctly.

Order of Operations	Symbol
1. Simplify grouping symbols.	(parentheses) [brackets] $\dfrac{2-1}{b-3}$ fraction bar
2. Find the values of all powers.	base → 2^{5} ← exponent
3. Multiply and **divide** in order from left to right.	× • ÷ /
4. Add and **subtract** in order from left to right.	+ −

VOCABULARY

base
   In a power, the number used as a factor; in 10^3, the base is 10. That is, $10^3 = 10 \times 10 \times 10$.

exponent
   In a power, the number of times the base is used as a factor; in 5^3, the exponent is 3. That is, $5^3 = 5 \times 5 \times 5$.

order of operations
   the rules that tell which operation to perform first when more than one operation is used

Copyright © Glencoe/McGraw-Hill, a division of The McGraw-Hill Companies, Inc.

Sometimes parentheses are used to set a number apart from other operations. If there is no operation to be performed inside the parentheses, check for **exponents**.

Example 1

Find the value of $4 - 2 + 16 \div 4$.

Use the order of operations. There are no grouping symbols or exponents.

$$4 - 2 + 16 \div 4 = 4 - 2 + 4 \quad \text{Multiply and divide from left to right.}$$
$$= 2 + 4 \quad \text{Add and subtract from left to right.}$$
$$= 6$$

From left to right, subtraction comes first in this expression.

YOUR TURN!

Find the value of $10 - 5 + 6 \cdot 3$.

Use the order of operations. There are no grouping symbols or exponents.

$$10 - 5 + 6 \cdot 3 = 10 - 5 + \underline{\hspace{2cm}}$$
$$= \underline{\hspace{1.5cm}} + \underline{\hspace{1.5cm}}$$
$$= \underline{\hspace{1.5cm}}$$

Example 2

Find the value of $71 - \dfrac{36 - 12}{6 + 2} \cdot 4^2$.

$71 - \dfrac{36 - 12}{6 + 2} \cdot 4^2 = 71 - \dfrac{24}{8} \cdot 4^2$ Simplify grouping symbols.

$= 71 - \dfrac{24}{8} \cdot 16$ Simplify exponents.

$= 71 - 3 \cdot 16$ Divide.

$= 71 - 48$ Multiply.

$= 23$ Subtract.

YOUR TURN!

Find the value of
$56 \div 14 + (1 + 4)^2 \cdot 2 - 4$.

$$56 \div 14 + (1 + 4)^2 \cdot 2 - 4$$

$$= 56 \div 14 + \underline{\hspace{1cm}} \cdot 2 - 4$$

$$= 56 \div 14 + \underline{\hspace{1cm}} \cdot 2 - 4$$

$$= \underline{\hspace{1cm}} + \underline{\hspace{1cm}} - 4$$

$$= \underline{\hspace{1cm}} - 4$$

$$= \underline{\hspace{1cm}}$$

Example 3

Write and simplify an expression to answer the question.

Seth arranged 24 chairs in rows of 6. He took 1 chair out of each row. Then, he placed 10 more chairs on the stage. How many chairs did Seth leave on the stage?

1. Translate each phrase.

Word Phrase	Math Meaning
4 rows	4
24 chairs in rows of 6	$24 \div 6$
1 chair from each row	-1
Ten more	$+10$

2. Write the expression.

   $[4 \cdot (24 \div 6 - 1)] + 10$

3. Simplify the expression.

   $[4 \cdot (4 - 1)] + 10 = 12 + 10 = 22$

Seth arranged 22 chairs.

YOUR TURN!

Write and simplify an expression to answer the question.

Susana bought 2 packs of 8 fruit bars. She gave 13 fruit bars away to her friends. Then she purchased 3 packs of granola bars with 6 bars in each pack. How many snacks does Susana have now?

1. Translate each phrase.

Word Phrase	Math Meaning
2 packs of 8	
Gave away 13	
Then she purchased	
3 packs of 6	

2. Write the expression.

   _____

3. Simplify the expression.

   $(\underline{\hspace{0.6cm}} - \underline{\hspace{0.6cm}}) + \underline{\hspace{0.6cm}} = \underline{\hspace{0.6cm}} + \underline{\hspace{0.6cm}} = \underline{\hspace{0.6cm}}$

Susana has _____ snack bars.

Copyright © Glencoe/McGraw-Hill, a division of The McGraw-Hill Companies, Inc.

GO ON

Who is Correct?

Find the value of $25 \div 5 + (8 - 4)^2 \cdot 2$.

Cedric

$25 \div 5 + (8 - 4)^2 \cdot 2$
$= 25 \div 5 + 4^2 \cdot 2$
$= 5 + 16 \cdot 2$
$= 21 \cdot 2$
$= 42$

Gracia

$25 \div 5 + (8 - 4)^2 \cdot 2$
$= 5 + (64 - 16) \cdot 2$
$= 5 + (48) \cdot 2$
$= 5 + 96$
$= 101$

Hannah

$25 \div 5 + (8 - 4)^2 \cdot 2$
$= 25 \div 5 + 4^2 \cdot 2$
$= 25 \div 5 + 16 \cdot 2$
$= 5 + 32$
$= 37$

Circle correct answer(s). Cross out incorrect answer(s).

 Guided Practice

Name the step that should be performed first in each expression.

1 $8 \cdot 3 + (30 - 3) \div 6^2$ _____

2 $17 \div 1 - (12 + 2) \cdot 2$ _____

3 $18 + 5^2 \div 5 + 4 \cdot 3$ _____

4 $8 + 17 \div 7 \cdot 5 - 6$ _____

Step by Step Practice

5 Find the value of $17 - 6 \cdot (3 - 2)^2 - 5 + 2$.

Step 1 Use the order of operations. Simplify the grouping symbols.

$$17 - 6 \cdot (3 - 2)^2 - 5 + 2 = 17 - 6 \cdot (\text{\_\_\_\_\_})^2 - 5 + 2$$

Step 2 Simplify the exponent.

$$17 - 6 \cdot 1^2 - 5 + 2 = 17 - 6 \cdot \text{\_\_\_\_\_} - 5 + 2$$

Step 3 Multiply and divide.

$$17 - 6 \cdot 1 - 5 + 2 = 17 - \text{\_\_\_\_\_} - 5 + 2$$

Step 4 Add and subtract.

$$17 - 6 - 5 + 2 = \text{\_\_\_\_\_} - 5 + 2$$
$$= \text{\_\_\_\_\_} + 2$$
$$= \text{\_\_\_\_\_}$$

Copyright © Glencoe/McGraw-Hill, a division of The McGraw-Hill Companies, Inc.

Find the value of each expression.

6 $64 \div 16 + (5 \cdot 2)^2 - 23 = 64 \div 16 + ($_____$)^2 - 23$

$$= 64 \div 16 + \text{_____} - 23$$

$$= \text{_____} + \text{_____} - 23$$

$$= \text{_____} - 23$$

$$= \text{_____}$$

7 $50 \div (9 + 1) \cdot 4 \div 2 = $ _____

8 $30 \div \dfrac{43 - 8}{3 + 4} \div 2 \cdot 12 = $ _____

9 $20 - 4^2 \div 4 \cdot 2 + (20 - 17) = $ _____

10 $(21 - 20)^2 \cdot 50 \div 5 - (72 \div 8) = $ _____

Step by Step Practice

11 **TRANSPORTATION** McArthur Community Center has 2 vans that hold 12 passengers each. They own 6 more minibuses that will hold 20 passengers each. How many passengers can the community center transport in all?

Problem-Solving Strategies
☐ Draw a diagram.
☐ Guess and check.
☐ Act it out.
☑ Write an expression.
☐ Work backward.

Understand Read the problem. Write what you know.

There are _____ vans with _____ passengers each and _____ minibuses with _____ passengers each.

Plan Pick a strategy. One strategy is to write and simplify an expression.

Solve Translate each phrase.

Word Phrase	2 vans of 12	More	6 buses of 20
Math Meaning			

Write and simplify the expression using the order of operations.

$$= \text{_____} + \text{_____}$$

$$= \text{_____} + \text{_____}$$

$$= \text{_____}$$

The community center can transport _____ passengers.

Check You can draw a picture to check your answer.

GO ON

Copyright © Glencoe/McGraw-Hill, a division of The McGraw-Hill Companies, Inc.

Write and simplify an expression to solve each problem.

12 **GARDENS**   Cierra likes to plant flowers. She planted 2 daffodils. She also planted 3 rows of 4 tulips. Five of the flowers were eaten by squirrels. How many flowers were left? Check off each step.

_____ Understand: I underlined key words.

_____ Plan: To solve the problem, I will _____.

_____ Solve: The answer is _____.

_____ Check: I checked my answer by _____.

13 **SUPPLIES**   Caine bought 3 packs of markers. Each pack had 5 markers. He gave 7 markers to his brother. Then he bought 2 more packs with 18 markers in each. How many markers does Caine have now?

Word Phrase	3 packs of 5	Gave away 7	More	2 packs of 18
Math Meaning				

14 **PHOTOGRAPHY**   Marcos was using his new digital camera at a family reunion. He took 6 pictures of each of his four aunts. He deleted 2 of the photos. Then he took 10 pictures of each of his 8 cousins. Finally, he took 4 photos that included his grandfather. How many photos are left on his camera?

Word Phrase	Math Meaning
6 pictures of 4 aunts	
deleted 2	
10 pictures of each of his 8 cousins	
4 photos of grandfather	

_____

_____

15 **Reflect**   Explain why $40 \div 4 + 6$ has a different value than $40 \div (4 + 6)$.

_____

_____

Copyright © Glencoe/McGraw-Hill, a division of The McGraw-Hill Companies, Inc.

Skills, Concepts, and Problem Solving

Name the step that should be performed first in each expression.

16 $5 \cdot 2 + (17 \div 1) - 22$ _____

17 $4 \cdot (2 - 6)^2 + 12 \div 3$ _____

18 $(6 - 2^2 \cdot 4) - 16 \div 2$ _____

19 $9 + (6 - 1 \cdot 14) \div 2^2$ _____

20 $3[(75 + 75) \cdot 3] - 25$ _____

21 $\dfrac{18 + 66}{35 - 14} \cdot 3 + 2$ _____

Find the value of each expression.

22 $14 - (7 + 5) + 7 \cdot 4^2 = 14 - \text{\_\_\_\_\_} + 7 \cdot 4^2$

$= 14 - \text{\_\_\_\_\_} + 7 \cdot \text{\_\_\_\_\_}$

$= 14 - \text{\_\_\_\_\_} + \text{\_\_\_\_\_}$

$= \text{\_\_\_\_\_} + \text{\_\_\_\_\_}$

$= \text{\_\_\_\_\_}$

23 $48 \div \dfrac{(37 + 3)}{(9 - 4)} - 4 \div 2 \cdot 7^2 = 48 \div \text{\_\_\_} - 4 \div 2 \cdot 7^2$

$= 48 \div \text{\_\_\_\_\_} - 4 \div 2 \cdot 7^2$

$= 48 \div \text{\_\_\_\_\_} - 4 \div 2 \cdot \text{\_\_\_\_\_}$

$= \text{\_\_\_\_\_} - \text{\_\_\_\_\_} \cdot \text{\_\_\_\_\_}$

$= \text{\_\_\_\_\_} - \text{\_\_\_\_\_}$

$= \text{\_\_\_\_\_}$

24 $50 \div 5 + 3 \cdot 2^2 - (15 - 9) = \text{\_\_\_\_\_}$

25 $3^2 + 8 \div 2 - (10 + 2) = \text{\_\_\_\_\_}$

26 $18 - 5^2 \cdot 0 + 16 - 15 = \text{\_\_\_\_\_}$

27 $(9 - 6)^2 + 8 \div 4 + 5 \cdot 6 = \text{\_\_\_\_\_}$

28 $\dfrac{27 + 23}{16 + 9} \cdot 5 = \text{\_\_\_\_\_}$

29 $\dfrac{4^2}{2 + (27 \cdot 0)} = \text{\_\_\_\_\_}$

30 $10[8(2^2 + 2) - (2 \cdot 6)] = \text{\_\_\_\_\_}$

31 $5[(17 \cdot 1) - 3(25 \div 5)] = \text{\_\_\_\_\_}$

Copyright © Glencoe/McGraw-Hill, a division of The McGraw-Hill Companies, Inc.

GO ON

Write and simplify an expression to solve each problem.

32 **COLLECTIONS** Evan had 100 bobble heads. He sold 5 sets of 10 bobble heads. He then bought 3 sets of 12 bobble heads. Then Evan sold 25 bobble heads. How many bobble heads does Evan have left?

Word Phrase	Math Meaning
had 100	
sold	
5 sets of 10	
bought	
3 sets of 12	
sold 25	

_____

_____

33 **BOOKS** Each week Serena uses her library card. On her first visit she borrowed 2 stacks of 8 books. She returned 9 books on the second week. On the third week, Serena borrowed 2 stacks of 5 books. How many books does Serena have now?

Word Phrase	2 stacks of 8	returned 9	2 stacks of 5
Math Meaning			

_____

Vocabulary Check **Write the vocabulary word that completes each sentence.**

34 The_____ is a set of rules that tells what order to follow when evaluating an expression.

35 A(n)_____ is the number of times a base is multiplied by itself.

36 **Writing in Math** Does $40 - (7 - 5)$ equal $(40 - 7) - 5$? Explain.

_____

_____

STOP

Copyright © Glencoe/McGraw-Hill, a division of The McGraw-Hill Companies, Inc.

Evaluate Algebraic Expressions

KEY Concept

To **evaluate** an **algebraic expression**, substitute a value for a variable. Then perform the operations.

$n = 2$ $p = -3$

$$4n + 5p = 4(2) + 5(-3)$$
$$= 8 + (-15)$$
$$= -7$$

Remember to use the **order of operations** after substituting, or replacing, the variables with numbers.

VOCABULARY

algebraic expression
a combination of numbers, variables, and at least one operation

evaluate
to find the *value* of an *algebraic expression* by replacing variables with numbers

order of operations
the rules that tell which operation to perform first when more than one operation is used
1. Simplify the expressions inside grouping symbols, like parentheses.
2. Find the value of all powers.
3. Multiply and divide in order from left to right.
4. Add and subtract in order from left to right.

Example 1

Evaluate ▲ + (9 − 2) + □ when ▲ = 3 and □ = 1.

1. Replace ▲ with 3 in the expression.

   ▲ + (9 − 2) + □ = 3 + (9 − 2) + □

2. Replace □ with 1 in the expression.

   3 + (9 − 2) + 1

3. Simplify. Follow the order of operations.

   3 + (9 − 2) + 1 = 3 + 7 + 1
                  = 11

YOUR TURN!

Evaluate 2 ◯² + 4 ♥ when ◯ = 2 and ♥ = 3.

1. Replace ◯ with 2 in the expression.

   _____

2. Replace ♥ with 3 in the expression.

   _____

3. Simplify. Follow the order of operations.

   _____

GO ON

Copyright © Glencoe/McGraw-Hill, a division of The McGraw-Hill Companies, Inc.

Example 2

Evaluate 4 ÷ y + x • 3 − 7 when x = 5 and y = 2.

1. Replace x with 5 and y with 2 in the expression.

$$4 ÷ y + x • 3 − 7 = 4 ÷ 2 + 5 • 3 − 7$$

2. Simplify using the order of operations.

$$4 ÷ 2 + 5 • 3 − 7$$

$= 2 + 5 • 3 − 7$	Divide.
$= 2 + 15 − 7$	Multiply.
$= 17 − 7$	Add.
$= 10$	Subtract.

YOUR TURN!

Evaluate $3y^2 + x • 3 − 2$ when x = 4 and y = 2.

1. Replace y with 2 and x with 4. Write the expression.

_____

2. Simplify using the order of operations.
$3(2)^2 + 4 • 3 − 2$

$=$ _____ _____

$=$ _____ _____

$=$ _____ _____

$=$ _____ _____

Who is Correct?

Evaluate the expression 12x − 5 + 4y • 2 when x = 4 and y = 2.

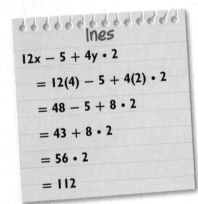

Ines

$12x − 5 + 4y • 2$

$= 12(4) − 5 + 4(2) • 2$

$= 48 − 5 + 8 • 2$

$= 43 + 8 • 2$

$= 56 • 2$

$= 112$

Sinclair

$12x − 5 + 4y • 2$

$= 12(4) − 5 + 4(2) • 2$

$= 48 − 5 + 8 • 2$

$= 48 − 5 + 16$

$= 59$

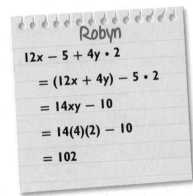

Robyn

$12x − 5 + 4y • 2$

$= (12x + 4y) − 5 • 2$

$= 14xy − 10$

$= 14(4)(2) − 10$

$= 102$

Circle correct answer(s). Cross out incorrect answer(s).

Guided Practice

Evaluate each expression when ☆ = 6.

1 $9 • ☆$ _____

2 $72 ÷ ☆$ _____

3 $4 + ☆ − 5 • 3 ÷ 3$

4 $8 • 4 + ☆ + 7$

_____

_____

Copyright © Glencoe/McGraw-Hill, a division of The McGraw-Hill Companies, Inc.

Evaluate each expression when ☺ = 8 and ♣ = 6.

5 ☺^{2} + 9 − 7 + ♣ · 10

Replace symbols with values: _____

Value of the expression: _____

6 10^2 ÷ 20 − (−6 + ♣) · ☺

Replace symbols with values: _____

Value of the expression: _____

7 $(27 − 18)^2$ + ☺ − 12 ÷ 4 + ♣ · 2

Replace symbols with values:

_____

Value of the expression: _____

8 16 ÷ 4 · ♣ − 2 + (☺ − 5)

Replace symbols with values:

_____

Value of the expression: _____

Step by Step Practice

9 Evaluate the expression $5y + 2z − 4$ when $y = 7$ and $z = 10$.

Step 1 $5y$ means 5 _____ y. Replace y with _____ in the expression.

Step 2 $2z$ means 2 _____ z. Replace z with _____ in the expression.

Step 3 Write the expression with all substitutions made.
Simplify using the order of operations.

$5 · 7 + 2 · 10 − 4 =$ _____ + _____ − _____

$=$ _____ − _____

$=$ _____

The value of the expression is _____.

Copyright © Glencoe/McGraw-Hill, a division of The McGraw-Hill Companies, Inc.

Evaluate each expression when $x = 2$ and $y = 5$.

10 $7y - (5 + 1) \div 2 \cdot x^2 = 7(\underline{\hspace{0.8cm}}) - (5 + 1) \div 2 \cdot (\underline{\hspace{0.8cm}})^2$

$$= 7(\underline{\hspace{0.8cm}}) - \underline{\hspace{0.8cm}} \div 2 \cdot 2^2$$

$$= 7(\underline{\hspace{0.8cm}}) - \underline{\hspace{0.8cm}} \div 2 \cdot \underline{\hspace{0.8cm}}$$

$$= \underline{\hspace{0.8cm}} - \underline{\hspace{0.8cm}} \cdot \underline{\hspace{0.8cm}}$$

$$= \underline{\hspace{0.8cm}} - \underline{\hspace{0.8cm}}$$

$$= \underline{\hspace{0.8cm}}$$

11 $5 - x \div 2 + (3 \cdot 2)^2 - 5 \cdot 0 = 5 - \underline{\hspace{0.8cm}} \div 2 + (3 \cdot 2)^2 - 5 \cdot 0$

$$= 5 - \underline{\hspace{0.8cm}} \div 2 + \underline{\hspace{0.8cm}}^2 - 5 \cdot 0$$

$$= 5 - \underline{\hspace{0.8cm}} \div 2 + \underline{\hspace{0.8cm}} - 5 \cdot 0$$

$$= 5 - \underline{\hspace{0.8cm}} + \underline{\hspace{0.8cm}} - \underline{\hspace{0.8cm}}$$

$$= \underline{\hspace{0.8cm}}$$

12 $16 + 4^2 \cdot x - 5 + (8 - y) - 0$

Replace variables with values:

_____

Value of the expression: _____

13 $5y^2 - 10 \div 5 + 3 \cdot 5x$

Replace variables with values:

_____

Value of the expression: _____

14 $(x^2 - 1) + 3 \cdot 4 \div (7 - 1) + y$

Replace variables with values:

_____

Value of the expression: _____

15 $100 \div y^2 + (x + 7)^2$

Replace variables with values:

_____

Value of the expression: _____

Copyright © Glencoe/McGraw-Hill, a division of The McGraw-Hill Companies, Inc.

Copyright © Glencoe/McGraw-Hill, a division of The McGraw-Hill Companies, Inc.

Step by Step **Problem-Solving Practice**

Solve.

Problem-Solving Strategies
☐ Draw a diagram.
☑ Use an equation.
☐ Guess and check.
☐ Act it out.
☐ Solve a simpler problem.

16 **BICYCLING**   It takes Larisa an hour to bicycle 12 miles. The total number of miles biked is b.

Use the expression $b \div 12$ to find how many hours it will take her to finish a trail ride. How long will it take for Larisa to complete a 60-mile trail?

Understand   Read the problem. Write what you know.

Larisa is completing a trail that is _____ miles long.

She bikes _____ miles each hour.

Plan   Pick a strategy. One strategy is to use an equation.

Use h to represent hours. Write an equation using h and the expression $b \div 12$.

$h = b \div 12$

Solve   In the equation, replace b with _____.

$h = \text{_____} \div 12$

Simplify.

$h = 60 \div 12$

$h = \text{_____}$

It will take Larisa _____ hours to complete the trail.

Check   Multiply to check your division.

17 **CLOTHES**   Shawnell wants to buy an $8 T-shirt and 3 sweaters. Use the variable expression $8 + 3s$ to find the total cost, where s represents the cost per sweater. Evaluate the expression for sweaters that cost $18 each. Check off each step.

_____ **Understand: I underlined key words.**

_____ **Plan: To solve the problem, I will** _____.

_____ **Solve: The answer is** _____.

_____ **Check: I checked my answer by** _____.

GO ON

Lesson 3-2 Evaluate Algebraic Expressions   **109**

18 FOOD Gabe's Grocery pays $26 per case for oranges. Write an expression for the cost of c cases. Find the cost of 8 cases. _____

19 Reflect Does the expression $50 \div k - 2$ have a greater value when $k = 5$ or $k = 10$? Explain.

_____

_____

▶ Skills, Concepts, and Problem Solving

Evaluate each expression when ◇ = 5 and ○ = 3.

20 $4 + ○ - ◇$ _____

21 $16 \cdot ○ + ◇$ _____

22 $2^2 - 6 + ◇ \cdot ○^2$

23 $15 \div ○ \cdot ◇ - 11 + 7$

_____

_____

Evaluate each expression when x = 9 and y = 3.

24 $18 \div x \cdot (10 + y - x)$ _____

25 $90 - x^2 + 6 \div y \cdot 2$ _____

26 $x^2 \div y + 7 \cdot 2 - (6 \cdot 1)$ _____

27 $(8 \cdot 1) + 17 \cdot (4y - x)$ _____

28 $y^2 \div y + (x + y) \cdot 1$ _____

29 $(2x - 1) + x^2$ _____

Solve.

30 GEOMETRY The area of a rectangle equals the expression $\ell \cdot w$, where ℓ represents the length and w represents the width. Evaluate the expression to find the area of the rectangle at the right. _____

$w = 4$

$\ell = 6$

31 RECREATION Lamar plays a math game in which whole numbers are worth 10 points, decimals are worth 15 points, and fractions are worth 20 points. The total score equals the expression $10w + 15d + 20f$, when w represents the number of whole numbers, d represents the number of decimals, and f represents the number of fractions. Find Lamar's score when $w = 7, f = 11,$ and $d = 15$.

_____

Copyright © Glencoe/McGraw-Hill, a division of The McGraw-Hill Companies, Inc.

32 The amount of a number is its _____.

33 Finding the value of an algebraic expression by replacing variables with numbers is called _____ the expression.

34 **Writing in Math**   Explain how to evaluate $r - 8 \cdot 2$ when $r = 30$.

_____

_____

 Spiral Review

Find the value of each expression. (Lesson 3-1, p. 98)

35 $56 \div \dfrac{(24 + 16)}{(21 - 16)} \cdot 6^2 \div 9 = 56 \div \underline{\quad} \cdot 6^2 \div 9$

$$= 56 \div \underline{\quad\quad} \cdot 6^2 \div 9$$

$$= 56 \div \underline{\quad\quad} \cdot \underline{\quad\quad} \div 9$$

$$= \underline{\quad\quad} \cdot \underline{\quad\quad} \div 9$$

$$= \underline{\quad\quad} \div 9$$

$$= \underline{\quad\quad}$$

36 $27 - (16 + 5) + 7 \cdot 3^2 = 27 - \underline{\quad\quad} + 7 \cdot 3^2$

$$= 27 - \underline{\quad\quad} + 7 \cdot \underline{\quad\quad}$$

$$= 27 - \underline{\quad\quad} + \underline{\quad\quad}$$

$$= \underline{\quad\quad} + \underline{\quad\quad}$$

$$= \underline{\quad\quad}$$

Solve.

37 **FOOD**   Imani was grocery shopping for the week. She bought 3 packs of each of 6 snack crackers. Then she bought 2 pieces of each of 4 different fruits. At the checkout counter she returned one pack of snack crackers. How many items did Imani purchase?

Word Phrase	Math Meaning
3 packs of 6 crackers	
2 pieces of 4 kinds of fruit	
returned one pack of crackers	

_____

Copyright © Glencoe/McGraw-Hill, a division of The McGraw-Hill Companies, Inc.

Progress Check 1 _(Lessons 3-1 and 3-2)

Name each operation that should be performed first.

1 $8 - 4 \cdot (7 + 4)^2 \div 2$ _____

2 $3^2 \cdot 2 - (12 \div 4) + 6$ _____

Find the value of each expression.

3 $18 + 2^2 \div 4 \cdot (5 - 2) + 7 =$ _____

4 $10 - (2 - 1)^2 + 16 \div 2 \cdot (1 + 1) =$ _____

5 $28 \div 2^2 \cdot 8 + 4 \div 2 =$ _____

6 $64 \div 4^2 \cdot 25 - (30 - 18) \div 4 =$ _____

Evaluate each expression when ▬ = 3 and ▲ = 5.

7 $8 + 9 \cdot ▬ - ▲$

_____

8 $6 \div ▬ + ▲ \cdot 3 - 2$

_____

Evaluate each expression when y = 8 and x = 2.

9 $19 - 2^2 - (6 + 2y) + 3y \div 2$

_____

10 $3y + x^2 - 32 \div 4 + (4 + 2 \cdot 3)$

_____

Solve.

11 **BASKETBALL** Tama made six 2-point shots and two 3-point shots. How many points did Tama score?

_____

12 **SHOPPING** Payton had 50 pencils. He sold 3 bags of pencils with 5 pencils each. He then bought 2 packs of pencils with 10 pencils each. Then Payton gave 20 pencils to his sister. How many pencils does Payton have left?

_____

_____

13 **UNIFORMS** The school band bought uniforms. See the cost of the uniform at right. Write an expression for the cost of *u* uniforms. Find the cost of 12 uniforms.

_____

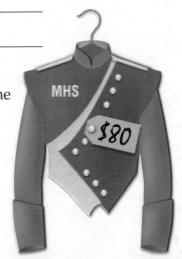

MHS

$80

Copyright © Glencoe/McGraw-Hill, a division of The McGraw-Hill Companies, Inc.

Copyright © Glencoe/McGraw-Hill, a division of The McGraw-Hill Companies, Inc.

Solve Algebraic Equations

KEY Concept

Variables can be used in **algebraic expressions** and **equations**.

Expressions	Equations
$5 + \square$	$10 \cdot \square = 20$
$8 - a$	$16 \div d = 4$

Inverse operations are opposite operations. They are used to undo each other. Addition and subtraction are inverse operations. Multiplication and division are also inverse operations.

You can use inverse operations to solve for x.

$$3 + x = 8, \text{ so } 8 - 3 = x \qquad x = 5$$

You can also think about a fact triangle to help you solve simple equations.

$3 + 5 = 8 \qquad 5 + 3 = 8$

$8 - 3 = 5 \qquad 8 - 5 = 3$

If you know that $3 + 5 = 8$, you can solve the equation $3 + x = 8$.

VOCABULARY

algebraic expression
a combination of variables, numbers, and at least one operation

equation
a mathematical sentence that contains an equals sign

inverse operations
operations that undo each other

variable
a symbol, usually a letter, used to represent a number

Fact triangles can help you find the unknown number. The equation shows the numbers 3 and 8. The number 5 is missing from the equation, so the value of x is 5.

Example 1

Find the value of the variable by modeling the equation $7 + \square = 12$.

1. Use the inverse operations of addition and subtraction.

   $7 + \square = 12$, so $12 - 7 = \square$ $12 - 7 = 5$, so $\square = 5$

2. Use a model to check your answer.
   Think: What number added to 7 equals 12?

   $7 + 5 = 12$ The value of $\square = 5$.

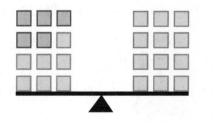

GO ON

YOUR TURN!

Find the value of the variable by modeling the equation 7 + ☐ = 15.

1. Use the inverse operations of addition
   and _____.

2. Use a model to check your answer.
   Think: What number added to 7 equals 15?

   $7 + \boxed{} = 15$

   The value of ☐ = _____.

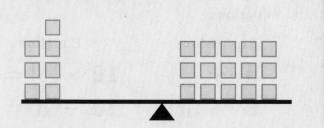

Example 2

Find the value of c in the equation 7 · c = 56.

1. Use the inverse operations of
   multiplication and division.

   $7 \cdot c = 56$, so $c = 56 \div 7$

   $56 \div 7 = 8$, so $c = 8$

2. Check your answer by substituting 8 for c.

   $7 \cdot c = 56$

   $7 \cdot 8 = 56$

   $56 = 56$ ✔

YOUR TURN!

Find the value of r in the equation r · 12 = 60.

1. Use the inverse operations of
   multiplication and division.

   $r \cdot 12 = 60$, so _____ = 60 ÷ 12

   60 ÷ 12 = _____, so _____ = r

2. Check your answer by substituting
   _____ for r.

   $r \cdot 12 = 60$

   _____ · 12 = 60

   _____ = 60

Copyright © Glencoe/McGraw-Hill, a division of The McGraw-Hill Companies, Inc.

Who is Correct?

Find the value of the variable in the equation 24 ÷ t = 3.

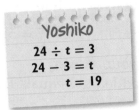

Dario
24 ÷ t = 3
24 ÷ 3 = t
t = 8

Pearl
24 ÷ t = 3
24 · 3 = t
t = 72

Yoshiko
24 ÷ t = 3
24 − 3 = t
t = 19

Circle correct answer(s). Cross out incorrect answer(s).

Guided Practice

Find the value of each variable by modeling the equation.

1 $5 + \square = 11$

$\square =$ _____

2 $7 - q = 3$

$q =$ _____

3 Find the value of f in the equation $\dfrac{f}{11} = 10$.

Step 1 $\dfrac{f}{11}$ means $f \div 11$. Use the inverse operations of multiplication and division.

$f \div 11 =$ _____, so $f = 11 \cdot$ _____

$11 \cdot$ _____ $=$ _____, so $f =$ _____

Step 2 Check your answer by substituting _____ for f.

$\dfrac{f}{11} = 10$

$\dfrac{\boxed{}}{11} = 10$

_____ $= 10$ ✔

Find the value of the variable in each equation.

4 $\square - 79 = 8$

$\square - 79 = 8$, so $\square = 8 +$ _____

$8 +$ _____ $=$ _____, so _____ $= \square$

5 $13 \cdot m = 52$

$13 \cdot m = 52$, so $m = 52 \div$ _____

$52 \div$ _____ $=$ _____, so _____ $= m$

6 $\dfrac{s}{6} = 7$

$s =$ _____

7 $100 + p = 143$

$p =$ _____

GO ON

Copyright © Glencoe/McGraw-Hill, a division of The McGraw-Hill Companies, Inc.

Step by Step Problem-Solving Practice

Solve.

Problem-Solving Strategies
- ☐ Draw a diagram.
- ☐ Guess and check.
- ☐ Use a model.
- ☐ Solve a simpler problem.
- ☑ Write an equation.

8 FINANCE Ms. Cartright had $47 in her wallet. She bought a birthday present for her best friend. Now she has $29 in her wallet.

How much money did she spend on the gift?

Understand Read the problem. Write what you know.

Ms. Cartright had _____.

She has _____ left.

Plan Pick a strategy. One strategy is to write and solve an equation.

Solve Let p represent the amount of money spent on the present. Write and solve the equation.

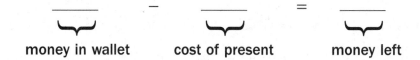

$$\underbrace{}_{\text{money in wallet}} - \underbrace{}_{\text{cost of present}} = \underbrace{}_{\text{money left}}$$

To solve the equation, use another subtraction sentence from the same fact family.

If _____ − _____ = _____,

then _____.

$p =$ _____

Ms. Cartright spent _____.

Check Substitute _____ for p in the equation.

47 − _____ = _____

_____ = _____ ✔

Copyright © Glencoe/McGraw-Hill, a division of The McGraw-Hill Companies, Inc.

Solve.

9 SHIPPING Dustin is packaging stuffed animals for a toy company. The shipping boxes will hold 14 toys each. How many boxes will he need to package 154 toys? Write an equation and solve for the variable. Check off each step.

_____ **Understand: I underlined key words.**

_____ **Plan: To solve this problem, I will** _____.

_____ **Solve: The answer is** _____.

_____ **Check: To checked my answer by** _____.

10 EXERCISE Mr. Castillo jogs 3 miles every day. How many days will it take him to jog 42 miles? Write an equation and solve for the variable.

_____

11 **Reflect** Explain how to use the fact triangle to solve the equation $x + 4 = 7$.

_____

_____

_____

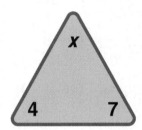

 Skills, Concepts, and Problem Solving

Find the value of each variable by modeling the equation.

12 $3 \cdot j = 15$

13 $4 + \square = 11$

$j =$ _____

$\square =$ _____

Copyright © Glencoe/McGraw-Hill, a division of The McGraw-Hill Companies, Inc.

GO ON

David Young-Wolff/PhotoEdit

Find the value of the variable in each equation.

14 $13 + n = 21$

$n = $ _____

15 $126 - \square = 106$

$\square = $ _____

16 $14 \cdot \square = 112$

$\square = $ _____

17 $\dfrac{240}{t} = 12$

$t = $ _____

Solve.

18 **ADVERTISING** Mr. Michaels is opening a new coffee shop. He printed 375 flyers to advertise his new shop. He has distributed 254 of the flyers. How many does he have left? Write an equation and solve for the variable

_____

19 **FINANCE** Juan Carlos earns $15 an hour. Last week he earned $330. How many hours did Juan Carlos work last week? Write an equation and solve for the variable.

_____

Vocabulary Check **Write the vocabulary word that completes each sentence.**

20 A _____ is a letter or symbol used to represent an unknown quantity.

21 **Writing in Math** Explain the difference between an algebraic expression and an equation.

_____

_____

 Spiral Review

Evaluate the expression when $\bigcirc$ **= 4 and** $\blacklozenge$ **= 3.** (Lesson 3-2, p. 105)

22 $\bigcirc^2 + 5 - 2 + \blacklozenge \cdot 8$

Replace symbols with values: _____

Value of the expression: _____

Copyright © Glencoe/McGraw-Hill, a division of The McGraw-Hill Companies, Inc.

Relate Algebraic Equations and Formulas

KEY Concept

Like other algebraic equations, many **formulas** use **variables** to show the relationships between values. Consider the algebraic equation $p + q = r$. The equation can be solved if the value of two variables is known. If $p = 3$ and $q = 2$, the value of r is 5.

$$p + q = r$$
$$3 + 2 = 5$$

Formulas often use more than one variable. For instance, distance, time, and rate are related values shown by the formula $d = r \cdot t$.

You can rewrite the formula to find the value of the unknown variable. Consider the following fact family.

$$d = r \cdot t \qquad d = t \cdot r$$
$$r = d \div t \qquad t = d \div r$$

Tanya traveled 50 miles in 2 hours. Substitute values for the variables to solve the equation.

$d = 50$ miles $\qquad t = 2$ hours $\qquad r = ?$

$r = d \div t$	Use the formula.
$r = 50 \div 2$	Substitute values.
$r = 25$	Solve.

The rate is 25 miles per hour.

VOCABULARY

area
the number of square units needed to cover the surface enclosed by a geometric figure

equation
a mathematical sentence that contains an equals sign

formula
an equation that shows a relationship among certain quantities

variable
a symbol, usually a letter, used to represent a number

Example 1

Use the formula $A = \ell \cdot w$ to solve for ℓ, length.

The area of the rectangle is 48 square centimeters. Its width is 8 centimeters. What is the length of the rectangle?

1. Substitute the values. $48 = \ell \cdot 8$

2. Use the inverse operation. $\dfrac{48}{8} = \dfrac{\ell \cdot 8}{8}$

3. What is the value of ℓ? $6 = \ell$

4. The length of the rectangle is 6 centimeters.

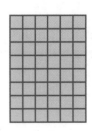

GO ON

Copyright © Glencoe/McGraw-Hill, a division of The McGraw-Hill Companies, Inc.

Copyright © Glencoe/McGraw-Hill, a division of The McGraw-Hill Companies, Inc.

YOUR TURN!

Use the formula $A = \ell \cdot w$ to solve for _w_, width.

The area of the rectangle is 35 square yards. Its length is 5 yards. What is the width of the rectangle?

1. Substitute the values.

$$\underline{\hspace{2cm}} = \underline{\hspace{2cm}} \cdot w$$

2. Use the inverse operation.

3. What is the value of _w_?

$$\underline{\hspace{2cm}} = w$$

Example 2

Use the formula $d = r \cdot t$ to solve for _r_, rate.

Shannon bicycled 36 miles in 3 hours. What was her rate of speed?

1. Substitute the values.

$36 = r \cdot 3$

2. Use the inverse operation.

$$\frac{36}{3} = \frac{r \cdot 3}{3}$$

3. What is the value of _r_?

$12 = r$

4. Shannon bicycled 12 miles per hour.

YOUR TURN!

Use the formula $d = r \cdot t$ to solve for _t_, time.

The Santa Lucia traveled 340 miles at 34 miles per hour. How many hours did the boat travel?

1. Substitute the values.

$$d = r \cdot t \qquad \underline{\hspace{1cm}} = \underline{\hspace{1cm}} \cdot t$$

2. Use the inverse operation.

3. What is the value of _t_?

$$\underline{\hspace{4cm}}$$

4. The boat traveled for $\underline{\hspace{2cm}}$ hours.

Who is Correct?

Find the value of f in the equation $a = c + f$, when $a = 47$ and $c = 38$.

Chenoa

$a = c + f$
$47 = 38 + f$
$47 + 38 = 38 + f$
$47 = f$

Shristi

$a = c + f$
$47 = 38 + f$
$47 - 38 = (38 + f) - 38$
$9 = f$

Damian

$a = c + f$
$38 = 47 + f$
$38 - 47 = (47 + f) - 47$
$-9 = f$

Circle correct answer(s). Cross out incorrect answer(s).

▶ Guided Practice

Find the value of h, when $b = 24$ and $e = 2$.

1 $b = e + h$

____ = ___ + ___

_____ = _____

$h =$ ____

2 $b = h - e$

____ = $h -$ ___

_____ = _____

$h =$ ____

Step by Step Practice

Use the formula $A = \ell \cdot w$ to solve for ℓ, length.

3 The area of the rectangle is 72 square meters. Its width is 12 centimeters. What is the length of the rectangle?

Step 1   Substitute the values.

_____ = $\ell \cdot$ _____

Step 2   Use the inverse operation.

$$\frac{\boxed{}}{\boxed{}} = \frac{\ell \cdot \boxed{}}{\boxed{}}$$

Step 3   What is the value of ℓ?

_____ = ℓ

Step 4   The length of the rectangle is _____ centimeters.

GO ON

Use the formula A = ℓ • w to solve for ℓ, length.

4 The area of the rectangle is 24 square meters. Its width is 6 meters. What is the length of the rectangle?

_____

5 The area of the rectangle is 32 square feet. Its width is 4 feet. What is the length of the rectangle?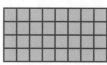

_____

Use the formula A = ℓ • w to solve for w, width.

6 The area of the rectangle is 21 square inches. Its length is 3 inches. What is the width of the rectangle?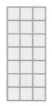

_____

7 The area of the rectangle is 25 square miles. Its length is 5 miles. What is the width of the rectangle?

_____

Step by Step Problem-Solving Practice

Use the formula d = r • t to solve for r, rate.

8 Tiffany traveled 34 miles in 2 hours. What was her rate of speed?

Understand Read the problem. Write what you know.

The distance, or _d_ is _____ miles.

The time, or _t_, is _____ hours.

Plan Pick a strategy. One strategy is to use a formula.

Use the formula $d = r \cdot t$.

Solve _____ = _r_ • _____   Substitute the variables for values.

$$\frac{\boxed{}}{\boxed{}} = \frac{r \cdot \boxed{}}{\boxed{}}$$ Use inverse operations.

_____ = _r_   Solve.

Tiffany's rate of speed is _____ miles per hour.

Check Substitute the values of _r_ and _t_ into the formula and solve for _d_.

Problem-Solving Strategies
☐ Draw a diagram.
☐ Use logical reasoning.
☐ Solve a simpler problem.
☐ Work backward.
☑ Use a formula.

Copyright © Glencoe/McGraw-Hill, a division of The McGraw-Hill Companies, Inc.

Use the formula $d = r \cdot t$ to solve for the missing variable.

9 **EXERCISE** Lazaro ran 15 miles at a rate of 5 miles per hour. How long did it take Lazaro to complete his run? Check off each step.

_____ Understand: I underlined key words.

_____ Plan: To solve this problem I will _____.

_____ Solve: The answer is _____.

_____ Check: I checked my answer by _____.

10 **SPACE TRAVEL** The space shuttle Atlantis travels at a rate of 17,500 miles per hour while in orbit. It has traveled 70,000 miles in orbit. How long has the shuttle been in orbit?

_____

11 **Reflect** Use the fact triangle to write four related multiplication and division equations for the formula of the area of a rectangle.

_____

▶ Skills, Concepts, and Problem Solving

Find the value of q, when $r = 37$ and $s = 19$.

12 $r = s + q$

_____ = _____ + _____

_____ = _____

$q =$ _____

13 $s = q - r$

_____ = $q -$ _____

_____ = _____

$q =$ _____

Use the formula $A = \ell \cdot w$ to solve for ℓ, length.

14 The area of the rectangle is 15 square kilometers. Its width is 5 kilometers. What is the length of the rectangle?

15 The area of the rectangle is 36 square millimeters. Its width is 6 millimeters. What is the length of the rectangle?

_____

_____

Copyright © Glencoe/McGraw-Hill, a division of The McGraw-Hill Companies, Inc.

Use the formula $A = \ell \cdot w$ to solve for w, width.

16 The area of the rectangle is 72 square meters. Its length is 9 meters. What is the width of the rectangle?

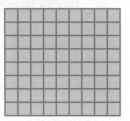

_____

17 The area of the rectangle is 28 square millimeters. Its length is 4 millimeters. What is the width of the rectangle?

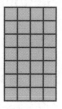

_____

18 The area of the rectangle is 49 square meters. Its length is 7 meters. What is the width of the rectangle?

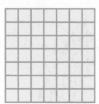

_____

19 The area of the rectangle is 30 square feet. Its length is 5 feet. What is the width of the rectangle?

_____

Use the formula $d = r \cdot t$ to solve for the missing variable.

20 **JETS**   The F-16 Fighting Falcon can travel 300 miles in 12 minutes. What is the jet's rate of speed?

_____

_____

21 **HELICOPTERS**   The Apache helicopter can travel 284 kilometers per hour. How long would it take the Apache to travel a distance of 1,420 kilometers at this rate of speed?

_____

22 **EXERCISE**   Maribelle entered a cross-country skiing race. She skiied at a rate of 8 miles per hour. If the race is 16 miles long, how long will it take Maribelle to complete the race?

_____

Copyright © Glencoe/McGraw-Hill, a division of The McGraw-Hill Companies, Inc.

Vocabulary Check **Write the vocabulary word that completes each sentence.**

23 A(n) _____ is a mathematical sentence that contains an equals sign.

24 An equation that shows a relationship among certain quantities is called a(n) _____.

25 _____ is the number of square units needed to cover the surface enclosed by a geometric figure.

26 **Writing in Math** The area of the rectangle to the right is 12 square units. Its width is 3 units. Find the length of the rectangle and explain how to use the diagram to check your answer.

_____

_____

 Spiral Review

Find the value of each variable by modeling the equation. (Lesson 3-3, p. 113)

27 $5 + h = 12$

28 $2 \cdot r = 18$

$h =$ _____

$r =$ _____

Find the value of the variable in each equation. (Lesson 3-3, p. 113)

29 $\square - 18 = 35$

$\square - 18 = 35$, so $\square = 35 +$ _____

$35 +$ _____ $= \square$, so _____ $= \square$

30 $8 \cdot g = 64$

$8 \cdot g = 64$, so $g = 64 \div$ _____

$64 \div$ _____ $=$ _____, so _____ $= g$

Solve. (Lesson 3-3, p. 113)

31 **ENTERTAINMENT** Marina and Dylan went to a baseball game. They bought snacks that cost $9.50. The total cost of the game tickets and the snacks was $27.50. How much did they pay for the tickets?

_____

STOP

Copyright © Glencoe/McGraw-Hill, a division of The McGraw-Hill Companies, Inc.

Progress Check 2 (Lessons 3-3 and 3-4)

Find the value of each variable by modeling the equation.

1 $9 - n = 4$ $n =$ _____

2 $2 + \square = 10$ $\square =$ _____

Find the value of the variable in each equation.

3 $15 + \square = 34$ $\square =$ _____

4 $50 - \square = 13$ $\square =$ _____

5 $9x = 108$ $x =$ _____

6 $\frac{m}{4} = 7$ $m =$ _____

Use the formula A = ℓ • w to solve for w, width.

7 The area of the rectangle is 54 square inches. Its length is 6 inches. What is the width of the rectangle?

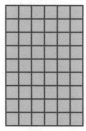

_____

8 The area of the rectangle is 49 square miles. Its length is 7 miles. What is the width of the rectangle?

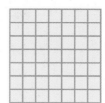

_____

Write an equation to represent each situation. Then answer the question.

9 **MONEY**   Jade earned $10 per hour last week. Her total earnings were $250. How many hours did Jade work last week?

_____

Use the formula d = r • t to solve for the missing variable.

10 **EXERCISE**   Lydia entered a walk for charity with her family. She walked an 8-mile course in 2 hours. If Lydia walked at the same rate for all 8 miles, what was her rate of speed?

_____

Copyright © Glencoe/McGraw-Hill, a division of The McGraw-Hill Companies, Inc.

Study Guide

Vocabulary and Concept Check

algebraic expression, *p. 105*

area, *p. 119*

base, *p. 98*

equation, *p. 113*

evaluate, *p. 105*

exponent, *p. 98*

formula, *p. 119*

inverse operations, *p. 113*

order of operations, *p. 98*

variable, *p. 113*

Copyright © Glencoe/McGraw-Hill, a division of The McGraw-Hill Companies, Inc.

Write the vocabulary word that completes each sentence.

1 To find the value of an algebraic expression by replacing variables with numbers is to _____ the expression.

2 $7x + 9 - 3y$ is an example of a(n) _____.

3 A symbol, usually a letter, used to represent a number is called a(n) _____.

4 Addition and subtraction are _____ because they undo each other.

5 _____
↓
$3x = 4 + 20$

6 The formula for _____ is $A = \ell \cdot w$.

7 In the _____, $d = r \cdot t$, the variable r represents the rate.

Write the correct vocabulary term in each blank.

8 _____
↓

9 _____
↓
2^5

10 _____

1. Simplify within parentheses.

2. Simplify exponents.

3. Multiply and divide from left to right.

4. Add and subtract from left to right.

3-1 Order of Operations (pp. 98–104)

Find the value of each expression.

11 $10 + 28 \div 7 - 3 \cdot 1 = $ _____

12 $13 - 9 + 12 \cdot 2 = $ _____

13 $5^2 + 20 \div 2 = $ _____

14 $172 + 12 \div 4 - 6^2 = $ _____

Find the value of each expression.

15 $6 \cdot 2 - (2^2 \cdot 2) + 21 \div 7 = $ _____

16 $4^2 \div (3 + 1) - 0 = $ _____

17 $(12 - 6)^2 + 3 \cdot 7 = $ _____

18 $(14 \div 2)^2 + 70 \div 10 = $ _____

Example 1

Find the value of $8 - 4 + 35 \div 7 \cdot 4$.

1. Use the order of operations.

2. There are no grouping symbols or exponents.

3. Multiply and divide.

$$8 - 4 + 35 \div 7 \cdot 4 = 8 - 4 + 5 \cdot 4$$
$$= 8 - 4 + 20$$

4. Add and subtract.

$$8 - 4 + 20 = 4 + 20$$
$$= 24$$

Example 2

Find the value of $18 \div 3 + (2 + 1)^2 \cdot 4 - 5$.

1. Use the order of operations.

2. Simplify the grouping symbols.

$$18 \div 3 + (2 + 1)^2 \cdot 4 - 5 = 18 \div 3 + 3^2 \cdot 4 - 5$$

3. Simplify the exponent.

$$18 \div 3 + 3^2 \cdot 4 - 5 = 18 \div 3 + 9 \cdot 4 - 5$$

4. Multiply and divide.

$$18 \div 3 + 9 \cdot 4 - 5 = 6 + 36 - 5$$

5. Add and subtract.

$$6 + 36 - 5 = 42 - 5$$
$$= 37$$

Copyright © Glencoe/McGraw-Hill, a division of The McGraw-Hill Companies, Inc.

3-2 Evaluate Variable Expressions (pp. 105–111)

Evaluate each expression when ☺ = 12.

19 $7 \cdot ☺$

_____

20 $120 \div ☺$

_____

21 $14 + ☺ \div 3 \cdot 5 - 1$

_____

22 $8 \cdot 4 - 3^2 + ☺ + 7$

_____

Evaluate each expression when $x = 4$ and $y = 0$.

23 $15 + 3^2 \cdot 4 - (x - 1) + 7y$

_____

24 $(2x^2 - 2) \div 5 + 4xy$

_____

25 $(5y + 10) \div (x - 2) + (3 \cdot 4)$

_____

26 $(x - 3)^2 + (15 - 4y)$

_____

27 $x \cdot y \cdot 3 + x \cdot 3 \div 6$

_____

Example 3

Evaluate $37 - \blacklozenge$ when $\blacklozenge = 24$.

1. Replace $\blacklozenge$ with 24 in the expression.

   $37 - \blacklozenge = 37 - 24$
2. Simplify. Follow the order of operations.

   $37 - 24 = 13$

Example 4

Evaluate $16 \div 8 + y \cdot x - 7$ when $x = 2$ and $y = 7$.

1. Replace x with 2 and y with 7 in the expression.

   $16 \div 8 + y \cdot x - 7 = 16 \div 8 + 7 \cdot 2 - 7$
2. Simplify using the order of operations.

Divide.	$= 16 \div 8 + 7 \cdot 2 - 7$
Multiply.	$= 2 + 7 \cdot 2 - 7$
Add.	$= 2 + 14 - 7$
Subtract.	$= 16 - 7$
	$= 9$

Copyright © Glencoe/McGraw-Hill, a division of The McGraw-Hill Companies, Inc.

3-3 Operations with Unknown Quantities (pp. 113–118)

Find the value of the variable in each equation.

28 $27 + x = 35$

$x = $ _____

29 $42 - b = 8$

$b = $ _____

30 $y + 32 = 72$

$y = $ _____

31 $m - 17 = 28$

$m = $ _____

Find the value of the variable in each equation.

32 $\dfrac{y}{7} = 8$

$y = $ _____

33 $f \cdot 7 = 63$

$f = $ _____

34 $\dfrac{t}{6} = 12$

$t = $ _____

35 $b \cdot 8 = 64$

$b = $ _____

Example 5

Find the value of a in the equation $a - 3 = 5$.

1. Use the fact that addition and subtraction are inverse operations.

   $a - 3 = 5$, so $a = 3 + 5$

   $3 + 5 = 8$, so $8 = a$

2. Use a model to check your answer.

3. Think: What number minus 3 equals 5?

   $8 - 3 = 5$

4. The value of a must be 8.

Example 6

Find the value of d in the equation $7 \cdot d = 35$.

1. Think: What number times 7 equals 35?

2. Use the fact that multiplication and division are inverse operations.

   $7 \cdot d = 35$, so $d = 35 \div 7$

   $35 \div 7 = 5$, so $5 = d$

3. Check your answer by substituting 5 for d.

   $7 \cdot d = 35$

   $7 \cdot 5 = 35$

   $35 = 35$

4. $35 = 35$ is a true statement, so 5 is correct.

Copyright © Glencoe/McGraw-Hill, a division of The McGraw-Hill Companies, Inc.

3-4 Relate Algebraic Equations and Formulas (pp. 119–126)

Use the formula $A = \ell \cdot w$ to solve for the missing variable.

36 The area of the rectangle is 16 square meters. Its width is 4 meters. What is the length of the rectangle?

_____

37 The area of the rectangle is 27 square meters. Its width is 3 meters. What is the length of the rectangle?

_____

Use the formula $d = r \cdot t$ to solve for the missing variable.

38 Jajuan can bicycle 14 miles per hour. How long would it take Jajuan to bicycle 42 miles at this rate?

_____

_____

39 A certain cargo plane traveled 1,000 miles in 5 hours. If the airplane traveled at the same rate for 5 hours, what was its rate of speed?

_____

_____

Example 7

Use the formula $A = \ell \cdot w$ to solve for ℓ, length.

The area of the rectangle is 30 square feet. Its width is 6 feet. What is the length of the rectangle?

1. Substitute the values.

$$A = \ell \cdot w \quad 30 = \ell \cdot 6$$

2. Use the inverse operation.
$$\frac{30}{6} = \ell \cdot \frac{6}{6}$$
$$5 = \ell$$

3. The length of the rectangle is 5 feet.

Example 8

Use the formula $d = r \cdot t$ to solve for t, time.

Janet's remote control car can travel 88 feet in one minute. How long would it take the car to travel 352 feet?

1. Substitute the values.

$$d = r \cdot t \qquad\qquad 352 = 88 \cdot t$$

2. Use the inverse operation.
$$\frac{352}{88} = \frac{88 \cdot t}{88}$$

3. The value of t is 4.

4. The car would travel for 4 minutes.

Copyright © Glencoe/McGraw-Hill, a division of The McGraw-Hill Companies, Inc.

Chapter Test

Find the value of each expression.

1 $16 + 4^2 \div 8 \cdot 5 - (16 - 10) =$ _____

2 $27 - 6^2 \div 9 + 5 - (5 \cdot 2) =$ _____

3 $(7 - 3)^2 \cdot 3 \div 4 + 11 + 8 =$ _____

4 $(5 - 2)^2 + 8 \cdot 5 - (27 \div 9) =$ _____

Evaluate each expression when $b = 4$ and $f = 2$.

5 $11 \cdot 3 + (b - f)^2 - 7 =$ _____

6 $b \div 2 + 7^2 + 5 - (5 \cdot f) =$ _____

7 $b^2 \cdot 3 \div b + 17 - f =$ _____

8 $7 + 4b \div f + (f + 3) - b^2 =$ _____

Find the value of the variable in each equation.

9 $27 + \square = 41$

$\square =$ _____

10 $56 \cdot z = 8$

$z =$ _____

11 $\dfrac{27}{n} = 3$

$n =$ _____

12 $p - 15 = 71$

$p =$ _____

Find the value of q, when $r = 37$ and $s = 19$.

13 $s = r + q$

____ = ____ + ____

_____ = _____

$q =$ _____

14 $r = q - s$

____ = $q -$ ____

_____ = _____

$q =$ _____

Use the formula $A = \ell \cdot w$ to solve for w, width.

15 The area of the rectangle is 44 square inches. Its length is 11 inches. What is the width of the rectangle?

_____

16 The area of the rectangle is 81 square miles. Its length is 9 miles. What is the width of the rectangle?

_____

Copyright © Glencoe/McGraw-Hill, a division of The McGraw-Hill Companies, Inc.

Solve. Explain your reasoning.

17 POPULATION An apartment complex has 4 units. Five people lived in each unit. Then 8 people moved away. The next month 2 families of 5 moved into the complex. What is the total number of people living in the apartment complex now?

_____

18 AGES Paulo is y years old. His aunt Serena is 17 years older. If his aunt is 34 years old, how old is Paulo?

_____

19 PRICES Each can of spaghetti sauce costs d dollars. If Erin buys 7 cans of sauce for $28, what is the cost of a can of spaghetti sauce?

_____

Use the formula $d = r \cdot t$ to solve for the missing variable.

20 EXERCISE Barkley went inline skating with his friends. If he travels 18 miles in 2 hours, what was his rate of speed?

_____

21 CARS During a 450-mile race, a certain car traveled at a rate of 150 miles per hour. How long did it take the car to finish the race?

_____

Correct the mistakes.

22 PHOTOGRAPHY Angelina's teacher asked, "If you have shipping boxes that will each hold 175 picture frames, then how many picture frames will x shipping boxes hold?" Angelina's answer is shown. What mistake did she make?

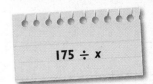

$175 \div x$

_____

_____

Copyright © Glencoe/McGraw-Hill, a division of The McGraw-Hill Companies, Inc.

Select the best answer and fill in the corresponding circle on the sheet at right.

1 $234 \div 3 \cdot [5^2 - (4 \cdot 3)] =$

   A 1,014

   B 954

   C 27

   D 6

2 What is the value of the expression?

$20 \div 5 + 17 \cdot (7 - 5)$

   A 36

   B 38

   C 42

   D 142

3 Evaluate $(5p - 2^2) \div 4$, if $p = 8$.

   A 9   **C** 144

   B 38   **D** 361

4 Evaluate $x^2 - 5y$, if $x = 15$ and $y = 9$.

   A −30   **C** 180

   B 75   **D** 1,980

5 What is the value of the expression?

$25 \cdot (5 - 2) \div 5 - 12$

   A −285   **C** 3

   B −3   **D** 62

6 Sasha's dad gave her $40 to take some friends to the movies. If movie tickets cost $8 per student, which equation will help Sasha figure out how many friends she can take? Let n equal the number of students going to the movies.

   A $\$8 - n = \$40$ **C** $\$8 \div n = \$40$

   B $\$8 \cdot n = \$40$ **D** $\$40 - n = \$8$

7 Laura and Eric baked 10 pies. Laura baked 3 pies. Which equation is used to find the number of pies Eric baked?

   A $3 + e = 10$ **C** $e - 3 = 10$

   B $3 \cdot e = 10$ **D** $10 \div e = 3$

8 Solve for m in the equation.

$\dfrac{48}{m} = 4$

   A 192   **C** 44

   B 52   **D** 12

9 Solve for ● in the equation.

$27 + ● = 52$

   A 25   **C** 1,404

   B 79   **D** 4,104

Copyright © Glencoe/McGraw-Hill, a division of The McGraw-Hill Companies, Inc.

10 Use the formula $d = r \cdot t$, to solve for r, rate.

Frank traveled 212 miles in 4 hours. What was his rate of speed?

A 212   C 50

B 53   D 4

---

11 The area of the rectangle is 42 square yards. Its width is 6 yards. What is the length of the rectangle?

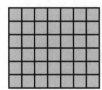

$$A = \ell \cdot w$$

A 252 yd   C 36 yd

B 48 yd   D 7 yd

---

12 Solve for q in the equation.

$$12 \cdot q = 60$$

A 5   C 48

B 6   D 72

Copyright © Glencoe/McGraw-Hill, a division of The McGraw-Hill Companies, Inc.

ANSWER SHEET

Directions: Fill in the circle of each correct answer.

1 Ⓐ Ⓑ Ⓒ Ⓓ
2 Ⓐ Ⓑ Ⓒ Ⓓ
3 Ⓐ Ⓑ Ⓒ Ⓓ
4 Ⓐ Ⓑ Ⓒ Ⓓ
5 Ⓐ Ⓑ Ⓒ Ⓓ
6 Ⓐ Ⓑ Ⓒ Ⓓ
7 Ⓐ Ⓑ Ⓒ Ⓓ
8 Ⓐ Ⓑ Ⓒ Ⓓ
9 Ⓐ Ⓑ Ⓒ Ⓓ
10 Ⓐ Ⓑ Ⓒ Ⓓ
11 Ⓐ Ⓑ Ⓒ Ⓓ
12 Ⓐ Ⓑ Ⓒ Ⓓ

Success Strategy

Read each problem carefully and look at each answer choice. Eliminate answers you know are wrong. This narrows your choices before solving the problem.

STOP

Copyright © Glencoe/McGraw-Hill, a division of The McGraw-Hill Companies, Inc.

Copyright © Glencoe/McGraw-Hill, a division of The McGraw-Hill Companies, Inc.

Success Strategy, 47, 95, 135

sum, 11

table
  function, 57
  make a, 54, 75, 84

term, 50

Test Practice, 46–47, 94–95, 134–135

variables, 113, 119

Vocabulary, 4, 11, 19, 27, 33, 50, 57, 64, 71, 79, 98, 105, 113, 119

Vocabulary Check, 10, 17, 26, 32, 38, 56, 62, 70, 78, 86, 104, 111, 118, 125

Vocabulary and Concept Check, 40, 88, 127

Who Is Correct?, 6, 13, 22, 29, 34, 52, 58, 65, 74, 82, 100, 106, 114, 121

whole numbers, 4

Writing in Math, 10, 17, 26, 32, 38, 56, 62, 70, 78, 86, 104, 111, 118, 125

x-axis, 64

x-coordinate, 64

x-values, 71

y-axis, 64

y-coordinate, 64

Your Turn, 4–5, 11–13, 20–22, 28–29, 33–34, 51–52, 57–58, 65, 72–73, 80–81, 98–99, 105–106, 114, 120

y-values, 71

Copyright © Glencoe/McGraw-Hill, a division of The McGraw-Hill Companies, Inc.